To Jeff & L
God bless you
THANK YOU for your help

Waves OF HOPE

The Impact of Galcom Radio Worldwide

Allan T. McGuirl
with
Florrie McGuirl

First Edition – July 2014

ISBN

978-1-4602-4213-1 (Hardcover)
978-1-4602-4214-8 (Paperback)
978-1-4602-4215-5 (eBook)

Cover Design by Dr. Shari Lou van de Pol

Cover Photograph by Ron Storer Photography

Edited by:
Bettina Von Kampen & Lois Neely Roberts

Produced by:
FriesenPress
Suite 300 – 852 Fort Street
Victoria, BC, Canada V8W 1H8
www.friesenpress.com

Distributed to the trade by The Ingram Book Company

GALCOM INTERNATIONAL
115 Nebo Rd., Hamilton, ON, Canada L8W 2E1
905-574-4626 or 1-877-242-5266
galcom@galcom.org | www.galcom.org

or in USA
Galcom International USA
P.O. Box 270956
Tampa, FL, USA 33688-0956
galcomusa@galcom.org

TABLE OF CONTENTS

DEDICATION

This book is dedicated to the glory of God
with thanks to my faithful, supportive wife, Florrie,
my constant encourager
to my children:
Allan David (Andrea), Ruth Anne, Betty Lynne (Peter),
Shari Lou (David) and Loralee (James)
who have worked alongside of me from the beginning,
to the many faithful financial and prayer supporters
and to my fellow founders,
Ken & Margie Crowell and Harold & Jo Ann Kent.

FOREWARD

I WAS ONLY A YOUNGSTER WHEN I FIRST MET ALLAN MCGUIRL; Allan and my grandfather pastored sister churches in Hamilton. So when Allan left the church to begin missionary work with Gospel Recordings (now Global Recordings Network) our church immediately started supporting Allan and Florrie.

I don't remember any other missionaries visiting our church, although I'm sure they did, but few people make an impression like Allan McGuirl. His cardboard record players, hand-wound cassette - all with funny sounding languages were fascinating to a young boy. Not to mention his woo-hooing and jumping around and getting us kids into the act! During the past five years of working and travelling with Allan I've learned that nothing about Allan is halfway, he does everything with 100% focus and energy.

But it's not just his focus and energy that set Allan McGuirl apart: it's that his energy is totally spent on service to Jesus Christ and his focus is completely on his Saviour. Everything Allan does, parenting, pastoring, leading a mission, eating in restaurants - everything_is done with a focus on Christ and a motivation to witness for Him. I cannot count the spiritual conversations I've watched Allan initiate with waitresses and hotel managers, or the number of people he offered to pray with when they mentioned some trouble they were facing. Allan seizes these "divine appointments" and it's amazing to watch how quickly that person is considering Christ.

"I'm just a piece of clay," Allan will say. And he's not just being humble, he's simply confessing his desire to be shaped by "The Master Potter", into a jar He can fill and use, a clay conduit to carry the life-giving, life-transforming Gospel.

As you read this book, it's our prayer that you will see what God can accomplish when an ordinary person gives himself totally to Jesus Christ. May you be challenged to make the same commitment Allan made to give every minute of your time, every part of your life, every ounce of your energy to the service of Jesus Christ.

Reverend Tim Whitehead,
Executive Director,
Galcom International

INTRODUCTION

IT IS AUGUST 31, 2012 AND I AM FLYING OVER THE NORTH Atlantic watching the shoreline of Greenland recede in the distance. I am 75, and have just spent twelve incredible days with a GALCOM team of nine people installing seven radio stations in remote areas of this vast, frigid country – that makes 125 stations installed globally and outreach into 160 countries. We left 1,000 "Go-Ye" solar-powered, fix-tuned radios with wonderful national pastors and a Danish missionary couple – we've prayed for twelve long years to watch God initiate radio ministry in this forgotten northland.

As I settle in for the flight back to Toronto, I'm thinking of the many amazing trips I've made for GALCOM since we started 25 years ago in August 1989: from freezing Greenland and Canada's Far North to the steaming jungles of Colombia and the D.R. Congo. From Burkina Faso to Bangalore. From Nicaragua to Nepal. Poland, Micronesia, Albania, Guatemala, Venezuela, India, Haiti, Paraguay - over 50 countries I've personally visited, and there are another 90 we've actively reached. All around the world the Gospel of Jesus Christ that changes lives has gone into homes and prisons through thousands of GALCOM's "Go-Ye" radio missionaries, carried in suitcases or dropped by parachute into remote and forbidden areas.

By the time this book is released Galcom will have sent out over a million little GO-YE radios and helped establish over 140 radio stations and penetrated over 160 countries around the world. The Lord is doing amazing work in country after country, nation after nation, language group after language group. Over and over again I have seen the results of the transforming power of the Gospel of Jesus Christ in cities, villages, jungles: lives changed from hopelessness to purposefulness, from evil to good, from sickness to health, from violence to peace, from deceitfulness to integrity, from darkness to light, from condemnation to eternal life.

Over 900,000 of the little Go-Ye radios have been placed in towns and villages around the world and now thousands of Envoy Players containing the Bible and/or messages have been shipped around the world. And we've just developed two exciting new products for delivering the gospel into extremely hostile areas, and helping train national pastors. – I've just taken the first new ImpaX audio player into Colombia, which carries a fix-tuned radio PLUS a Spanish Bible. And we've just sent 600 ImpaX players into Uganda via a local church missions group. A second high-tech product is letting us ship blank players into the most hostile countries; then via internet we can send the audio into the country to be loaded into the players for distribution. Hallelujah!

During these past 25 years we've seen firsthand the multiplied thousands of lives – and indeed entire communities that have been transformed by the simple message that *"God so loved the world that He gave his one and only Son that whoever believes in Him shall not perish but have eternal life."*

And I'm thinking of the many wonderful brothers and sisters in the Lord from the many tribes and languages who have blessed me with their joy and confident hope – this has been the best of all in my journeys. And I'm wondering: how did this little kid from the Ottawa Valley get involved in all this?

1

HE'S NOT GOING TO LIVE

"'For my thoughts are not your thoughts, neither are your ways my ways', saith the Lord." Isaiah 55:8

MAY 12TH, 1937 WAS A VERY SPECIAL DAY IN HISTORY, BRITISH history anyway, because that was the Coronation Day of King George VI, father of our present Queen. Over in Canada in a small village called Billings Bridge, just south of Ottawa our capital city, twin babies were born six weeks prematurely. Anne McGuirl came into the world first, followed by scrawny baby brother Allan. "He won't live out the day," the doctor warned my mother. But God's will was different from the doctor's prediction. With much care, and I think prayer, I began to thrive, although I remained small as a child. Mother already had two girls, Joan, twelve and Sheila, six when Anne and I were born. As twins born on this special day, my sister and I were presented with silver cups and bracelets inscribed with our names and the heads of the king and queen stamped in relief.

Our big brick 5-bedroom home had a wood stove and hand-pump at the sink, and a little building out back that served as toilet – visits were very short during the winter months. A creek ran through the back of our 10 acres and it was a great place for me and my three sisters to play.

Interestingly, my great grandfather, Patrick McGuirl was a Catholic priest in Ireland, but he fell in love with a young lady there. To resolve the dilemma, they married and immigrated to the Ottawa Valley where many Irish people had settled, and Patrick became a teacher. My grandfather, Thomas Henry McGuirl (1853 – 1914) settled in Billings Bridge in the early 1890's where my father, Allan Carleton McGuirl (1894-1940) was born at the family homestead. My father was first a school principal, then an immigration officer; he was also a musician with a dance band that often performed at the up-scale Chateau Laurier Hotel on the shore of the Ottawa River.

Sadness hit the home suddenly when at three years of age, my father had a massive heart attack and died instantly as he was walking out the side door. He had no life insurance in those days so Mother had to find work. What would she do with four children, the youngest 3 year old twins? Mom had used the services of a wonderful French-Canadian teenager from a poor family, Jean Villeneuve, barefoot and barely fifteen, who lived in with us as a nanny when we were first born. When Jean heard of my father's death, she volunteered to help out. What a wonderful person she was. As my mother worked long hours five and a half days a week, Jean, barely more than a child herself, kept the home front. All the years growing up, I give credit to my mother for the challenge she accepted in those days of raising a family while working full time in a "man's" world. She was among the first women tellers in the banking system in Ottawa often receiving surprised stares. She later became an accountant at the British Embassy and sometimes hosted visiting dignitaries.

I always enjoyed tinkering with things. My bedroom was something special. I could sit in my room with controls to open and close my door without getting out of bed. I wired alarms on my desk so my sisters could not borrow any of my pencils. I had several radios hooked up beside my bed and with an antenna outside my dormer window I could pick up programs from all over the world and had lots of fun listening into the night. Little did I know what God had planned for me in the future!

Our family occasionally attended the Presbyterian or United Church, and in my childhood we were taught to just live a good life and we would go to heaven. When my twin sister, Anne, and I were eight years old, an older couple from the Metropolitan Bible Church in Ottawa, Mr. & Mrs. Magee, sponsored us for a week at Bible camp. That was a life changing experience for my sister, Anne, who gave her heart to the Lord. I only gave "head consent" until age 13, when my best buddy, Bobby Booth and I were going to the Church of the Nazarene and we both won a week at camp for perfect attendance. On the Wednesday evening camp meeting the Lord spoke to both Bob and me and we went forward to give our lives to Christ. Jesus has been the major influence in my life ever since. Praise God!

While still in high school, I got a part time job at the Canadian Tire store in downtown Ottawa demonstrating wood working equipment. After graduation they moved me into the service department for small appliances and power tools, then, I became manager of the Adjusting Department - really the complaint department. I heard many wild stories about why certain items didn't work. Eventually, at age 27 I was put in charge of the Advertising and Public Relations for 23 stores throughout the Ottawa Valley.

During these years I attended the Metropolitan Bible Church (the "Met") where my sister Anne was very involved. I was baptized there and later worked with Open Air Campaigners with Len Percival.

My work took me several times to Canadian Tire headquarters in Toronto where I learned that a new store was being opened in Bells Corners just west of the city. In spring 1966 I was approached to be the store owner, and financial support began falling into place. These were rapidly expanding days and owing your own store almost guaranteed you would be set financially for life. I was just 29. Agreements were being drawn up for the dealership and I had to make a decision.

2

GOD IS CALLING

"Whether you turn to the right or to the left, your ears will hear a voice behind you, saying, 'This is the way; walk in it.'" Isaiah 30:21

A REAL STRUGGLE WAS GOING ON IN MY HEART BECAUSE since my mid-twenties onwards, I had sensed the Lord was calling me into the ministry. Then on August 6, 1966 at my office, I suddenly became ill. My whole right side was not functioning so I went to the hospital. Tests over several days all came back negative, puzzling the doctors.

One of the hospital cleaning ladies was from "The Met", and I asked her to contact Pastor Larson who arrived the following afternoon. After some small talk, since he knew me well, he bluntly said, "Allan, God wants you in the ministry." Then he showed me the total commitment of Psalm 37:5. "What are you going to do about it?" Larson challenged.

Right then and there I poured out my heart to the Lord. I surrendered my life totally willing to go anywhere, to do anything at anytime. A beautiful peace flooded my soul. Amazingly, the next morning whatever was affecting me was totally gone and has never bothered me since. There was no doubt where God was leading and the dealership faded into the background.

I conveyed my decision to the leadership of Canadian Tire and they accepted my decision graciously. I carried on working, waiting for the next step in God's call on my life.

In 1967, Canada's Centennial Year, I was a classic car buff and headed up the Antique Car Club of Ottawa. I owned several beauties, including an antique one-of-a-kind truck, and was driving a 1936 Ford. So I suppose it was only natural that I should be asked by head office to lead the Ottawa Valley portion of the "Tour Of Yesteryear".

This involved arranging for over 1,000 antique cars coming to Canada's capital for three days - what a massive undertaking it turned out to be! The parade of these beautiful antique cars drove right through the TV studios of CJOH with Bill Luxton interviewing different drivers. The cars proceeded past Parliament Hill then on to the Governor General's residence where I stood under the canopy beside Governor General Roland Michener for almost an hour as the parade passed by.

Allan with Governor General Roland Michener at Government House, 1967, for the Antique Car Event

The tour continued out to the Coliseum at Lansdowne Park where I had arranged a banquet for over 1200 people and prepared a program with special guests. Little did I realize how much this type of experience would benefit me in later years.

In April of that 1967 Centennial Year, a new Christian Education Director had come on staff at "The Met",

Miss Florrie Rout, a recent graduate of Toronto Bible College. Florrie was responsible for all the children and youth activities right up to College and Careers – the class I was in. I found her very charming and she appeared to be a godly person who really loved the Lord. But Florrie was steering clear of all the single young men as she gave great leadership.

To get some ministry experience, I started helping Pastor Bill Russell at Island Lodge, sometimes speaking to the seniors. Florrie often accompanied Pastor Russell to provide music.

Christmas was just around the corner and Florrie was getting ready for the children's Christmas concert. In early December, she came in to the Tuesday evening prayer meeting needing some help from the College and Career group to set up a large wall display at the front of the church over the baptistry. It so happened that everyone was busy except Gil McCormack and myself and I was delighted to be able to help Florrie. With ladders in place, we assembled the display and of course afterwards had to go for coffee. I asked Florrie if she was going to the Christmas Banquet, which she was, and offered to accompany her. Over the next few months, we met from time to time but were extremely careful to not infringe on her ministry responsibilities.

In early 1968, I was up in Toronto at the main Canadian Tire Store for meetings so took the opportunity to visit Toronto Bible College on Spadina Rd. Within thirty minutes I was accepted as a student for the next year. As I was leaving the Registrar's Office, I met Rev. Stan Beard who was Florrie's former pastor. When he heard I had registered for September, he offered me a place to stay. Wonderful! Praise God!

At the next Ottawa regional dealer's meeting, when I told them of my decision to resign, there was stunned silence. They had thought my idea of going into the ministry would pass. They could not believe I would leave a million dollar opportunity to go to Bible College! Things went smoothly for my departure. A man from the Cornwall store was prepared to step into my place so I could leave without concern. I had a buyer for my classic Oldsmobile and was offered a cheaper Volkswagen at a great price - four years of college would put me on a tight budget. At the farewell celebration for me at Canadian Tire, the dealers I had worked with over 15 years handed me an envelope. It contained just the right amount of money to cover my first year's tuition. Praise the Lord!

Just before leaving for Bible College, I made a very special proposal to a fine young lady named Florrie and we were engaged. Then I was off to Toronto for the first year of Bible College. By then, Toronto Bible College had merged with London Bible College to form Ontario Bible College. The following June 14, 1969 we were married and headed, right after our honeymoon, for a summer camp ministry together. Florrie had begun her teaching career at age 17 in a one-room school near Midland, Ontario, then she'd attended Toronto Bible College. During her summers she'd been Director of Children's Ministries at Fair Havens Bible Conference near Beaverton, Ontario. Before our wedding she had resigned from "The Met" and was preparing for another summer at Fair Havens. They also needed a Boys Camp Director so we spent our first weeks together in camp ministry.

WHERE GOD CALLS, HE PROVIDES

About three days before the end of camp, I received an amazing phone call. My twin sister Anne and her husband Bob had paid my full tuition for the next year of Bible College! All the way through College, God provided in various and unusual ways for our needs even when Allan David McGuirl was born on September 22, 1970. What a joy he was to us.

In April 1972, I graduated with a Bachelor of Religious Education degree with a pastoral major. This was a major challenge in itself as dyslexia, a lifelong struggle, threatened to scramble my mind constantly. With Florrie's tutoring my reading and research skills improved dramatically.

Just before graduation, in March, the professor of pastoral theology picked 10 men to meet with the pulpit committee from St. Marys, a small community near Stratford, Ontario. I remember as we waited to be called for the interview, looking at the other men and thinking, "What am I doing here? I'll never have a chance for this position." Well, the Lord thought differently and after I preached, I was given a unanimous call to become their first pastor. Accepting that call was definitely a step of faith as there were only six families in this new church who announced they were trusting the Lord to provide the funds to pay me. Florrie and I simply agreed: "If the Lord calls He will provide". And He did.

We bought a small house in June 1972 and began three and a half wonderful years of ministry in pioneering this new work. Soon we had a congregation of over 100. It was here, in Grace Community Church, on February 8, 1973 that I was ordained as a minister under the Associated Gospel Churches of Canada (the AGC).

The director of Fair Havens Bible Conference, Rev. Bill Crump, had become our dear friend. Bill was also serving as interim pastor of Mountain Gospel Church in Hamilton, and asked me if I could fill in for him some Sunday in March 1975. So Florrie and I visited Hamilton. I preached morning and evening and Florrie sang. The church had been through some difficult times and attendance was extremely low: about 36 people were in the Sunday morning service. The church building was run down and the people were hurting.

After the evening service a committee invited me into the Pastor's Study and asked me to consider a call. I remember telling them I was not looking for a change since we were very happy in St. Marys. But we would ask the Lord what He wanted us to do. A little over three weeks later we received a letter from this church that they had voted and it was 100% unanimous that I should come as pastor. This was a difficult challenge. The folks at St. Marys had become so precious to us. By now we had three children, Allan David was almost five, Ruth Anne was almost three and Betty Lynne was almost a year old. And we'd bought a house.

But Florrie and I had been praying, and somehow we knew this was a call from God. And where God calls, He provides. So we tearfully tendered our resignation – it's always hard for a congregation to understand this "call from God" that pastors sometimes feel. We sold our little house and in August 1975 moved into the Mountain Gospel Church parsonage.

It didn't take long to get settled in Hamilton. The first challenge was to knock off the $40,000 church debt. We worked at improving the building and started a Family Night. Florrie started an excellent choir and used her teaching and camping skills to increase our ministry to children and youth. The downtown Caroline Gospel Church merged with our congregation and the name was changed to Mountain Bible Church. God was blessing the ministry and soon the debt was retired.

Our involvement in missions grew as well. Following the example of Toronto's Peoples Church with Oswald Smith's commitment to foreign missions, we began bringing in missionary speakers and promoted Missions Conferences with Faith Promise Offerings. The church was prospering. The children's and youth ministries had grown so much that we needed to add a Christian Education Wing, and praise God, we had funds to build it!

A NEW DIRECTION

While pastor at Mountain Bible Church, I was also serving on the board of Gospel Recordings. The goal of this ministry is to record Scripture portions and messages in all the languages of the world – over 12,000 of them. These were put onto mini 78 RPMs and distributed among peoples who had no written language or were unable to read what was available. Florrie and I knew the work fairly well and rejoiced that they were getting the gospel message into hundreds of unreached language groups. For several years GR had been operating without a Canadian director and the ministry was going down fast. In December 1981 the Board Chairman, Bob Phillips, approached me about taking on the position of Canadian Director. We agreed to pray about it.

This was a hard one. The work of Mountain Bible Church was going so very well. The church had grown to almost full attendance on Sundays and over 200 attending our Tuesday Family Night program. Moreover, Gospel Recordings required their workers to step out in faith without deputation work, no visible means of support. By this time, Shari Lou and Loralee had been added to our family – that made seven McGuirls. We would have to give up the parsonage so we would have no place to live. This was the toughest decision Florrie and I were ever called on to make.

A second time the Gospel Recordings Board, unanimous in their decision, asked me to take the position. All that winter we thought and prayed about it. We sensed the Lord saying: *"This is the way, walk ye in it"*. We knew we would be in total dependence on the Lord to provide for us. In June 1982 I handed in my resignation at Mountain Bible Church effective August 1st. We said our goodbyes and took up the work of Gospel Recordings in August.

But where would the seven McGuirls live? God had plans for us. First, the church allowed us to stay at the parsonage until they found a new pastor. Then the miracles began as God opened up the windows of heaven, as it were, to provide a home for our family.

3

THE MIRACLE HOUSE

"But my God shall supply all your need according to his riches in glory by Christ Jesus." Phil. 4:19

AGAIN GOD HAD PLANS FOR US. JUST AFTER I HAD ANNOUNCED my resignation from Mountain Bible Church, Florrie and I went house hunting for a place for our family of seven. There was nothing in our price range. One day I was driving along Stone Church Road East: it was like a country road at that time. It was then I saw the house on the hill. I came home and told Flo, "I think I've found a place for us to live. The house isn't much but the view is terrific." My wife shares the beginning of this story:

"'The house isn't much but the view is terrific!' These were Allan's excited words as he began to describe the "house" on the hill. Months of fruitless searching just might be over.

"The 'house' or haunted house as the neighbourhood children called it, was quite a sight. The windows were boarded up, the lawn overgrown concealing junk and debris, stairs leading to the front door no longer existed, the lean-to was leaning precariously to the east and the southern portion of the roof had a definite concave appearance from ground level.

"On entering the house, a distinct mouldy odour assaulted the senses. Carefully picking our way across the kitchen floor revealed a chef's nightmare. Cupboards consisted of orange crates nailed to the walls with makeshift doors. One small window permitted enough light to recognize the remains of an ancient stove. The floor slanted decidedly toward the basement door. Aaah! Now there was something – an actual door. Further exploration of the house would mark this as an exception.

"Attached to the kitchen was the main bathroom where the toilet was recessed into the floor at an odd angle. Apparently, according to neighbours, the man who previously inhabited the house had propped up the toilet from the basement with a long pole to prevent it from making a grand entrance into the lower level. Considering that the house had been uninhabited for the previous seven years, it was surprising that the toilet was still there at all. The bathtub was a four-footed oddity the colour of which was questionable. I believe there was a sink there as well but that has faded from my memory. Throughout this investigation process my emotions were constantly shifting from shock to excitement to insanity.

The House on the Hill

"The small bedroom on the main floor was extremely dark. Oh, I forgot to mention that there was no electricity in the place. The broken, boarded up windows allowed precious little light to enter. Maybe that was a blessing. The living room and dining room were arched on either side of the main entrance. On first exploration, it escaped us that the living room ceiling had an interesting bow to it.

"This insul brick, two storey house had not been lived in for about 7 years. It had been up for sale by more than nine different real estate agencies, to which the signs in the basement attested! It had no glass in the windows so most were boarded up. There was running water – from the roof to the basement which created its own indoor murky pool of sorts, no furnace, no working plumbing, no electricity and for the most part, no doors. Kids had broken in and damaged walls. It needed major surgery. We contacted the real estate agent to find out the price thinking that we could work away at it and fix it up one room at a time. Since the real estate market was suffering and the owners were anxious to sell, we were told to make any ridiculous offer and he would see what he could do. Ultimately, we bought the derelict house for the price of the land. God even worked out the financial arrangements to suit our situation. We were now the proud owners of what would become 'God's miracle house'".

Then we watched God work. Our nephew Peter Rout was 19 at the time and had just finished high school. Uncertain about his future, when he heard about the house, he volunteered to come and help. He worked from July until October doing all kinds of jobs from completely gutting the house to figuring out how to make the renovations match the blue prints we had drawn up. His help was invaluable. Interestingly he now runs his own renovation business. His dad, Bill, also made several trips from Toronto to lend a hand.

Our brother-in-law, Bob Thompson, while reading the newspaper noticed a large house in North Toronto was being demolished. The buyer of the house liked the area but his wife didn't care for the house, quite posh in those days, even though it had been newly decorated. His secretary suggested that he sell off what he could before bulldozing the place. Bob went that night to check it out. He was able to buy three bathrooms – all fixtures including plumbing for $100 apiece. He called us that night to say what he had done hoping that we really would back up the offer. He also suggested I join him Saturday morning to see what else we might be able to use. We bought doors with frames for $10 to $20 each, long baseboard heaters for $10, rugs for $20, 200 amp electrical panel with all the wiring we could salvage for $100, stairs for $5, all the copper pipe and other fittings that we had time to remove for nothing, a closet full of drapes, not yet hung, for free. In all, we calculated that for about $1,200 we were able to purchase over $25,000 worth of goods. Bob and I worked hard for four days to remove and transport everything to Hamilton before the house was bulldozed.

Next, Bob noticed a place that sold windows up near Orangeville that was going bankrupt. We ended up getting all our triple glaze windows including two large picture windows for $50 to $100 each. For $700 we were able to purchase over $5,000 worth of windows. We also needed to add a small addition to the house and a friend of ours, John Durley, came to do the block work. John was almost finished and we were ready to start framing.

That Sunday, another pastor was coming to preach for a call at Mountain Bible Church, so we decided to take our children elsewhere. As circumstances worked out, we only had time to get to the large Pentecostal church nearby. Among the thousand plus attendees, we happened to meet a former member from our church, Harry Wiersma who was visiting that church for the day. When he found out we were moving into missions and were rebuilding a house, he volunteered to drop by Monday morning to see how things were progressing. True to his word, he showed up early the next morning. As a custom builder, he picked up the blueprints, sized up the situation and said, "As soon as the foundation is finished, give me a call. I have an apprentice who needs some experience and we will come and help." True to his word, when the foundation was complete, Harry showed up with his apprentice and got to work. I have never seen anything like it. In one fluid sweep, he would read the blueprint, pick up the lumber, saw it and nail it into place without a lost movement. The addition was up in no time - another amazing answer to prayer. Harry also remarked, "When you have the sonotubes in place for the front porch, I will come and frame it in so that the roof line doesn't have any leaks." He showed up a few weeks later and completed that job, too, at no charge.

We knew that we would have to repair the old roof and reshingle it along with shingling the new addition. Florrie's brother, Bob, came down to help for the day and remarked, "Oh, I had some roofing nails left over from doing my house so I brought them along" as he removed a keg of the long roofing nails from his trunk that we would need for the main part of the house. A few days later, Don and Barb Smith dropped in on their way back from Crystal Springs Camp. He had heard we were renovating a house and had a container of roofing nails. These were the short ones for the new roof and just enough for the addition. Even roofing nails, Lord!

We had a non-functional drilled well on the property so with a family of seven, we felt it necessary to hook up to the city water. We searched but could not find any main shut off at the road. We knew that the main water line was on the far side of the street. If there was no feed into our property, we would have to blast a trench across the solid rock road – a huge expense. Off I went to city hall to ask them to check it out. They never responded so I went a second time. This time they said their maps didn't even show a house on that property. Interestingly, the property tax department knew it was there! We assured them there was a house and they finally agreed to send a man to see if he could find the shut off. As the man started at the west end of the property tracing his detector over the ninety foot frontage of the lot, I was following behind praying, "Lord, help him find the valve. As he approached the final 8 feet of frontage, all of a sudden, "beep, beep, beep". Have you ever rejoiced over a "beep"? Well, we did. The man lifted his shovel and dug through the gravel: there just a few inches below the surface was the shut off valve. Praise the Lord!

Miracles kept on happening. The north block wall of the basement about four layers down was leaning out leaving about a 1 ½ inch gap between it and the wall below, probably the result of the continual freezing and thawing that took place over the seven years the house was vacant. We figured that when the digging was done for the weeping tiles, we would have to chip out each block and relay the upper section of the foundation. The back hoe was scheduled to arrive the Saturday of our Rout Family Reunion. We questioned whether we should stay home to supervise but felt the family reunion at Florrie's oldest brother Ron's cottage was too important to miss. When Monday morning arrived and Peter and I returned to the site, I no sooner had some tools out of the car and Peter called from the basement. I wondered what had happened now. Over the weekend, the leaning wall had fallen back up into place. As Peter's younger brother Tim surveyed the wall several days later he remarked, "I heard of the walls of Jericho falling down, but I've never heard of a wall falling up!" The crack in the wall was almost imperceptible so we just built a small retaining wall to support it.

That Monday, the back hoe returned to dig a trench from the house to the roadway. At that time, Stone Church Rd. was mainly a side road with ditches along both sides of the street. We would be able to direct the runoff water from the weeping tiles out to the ditch. About 15 feet out from the front of the house, the back hoe operator could go no farther. The rock was on an incline and therefore without blasting it would be impossible to complete the trench. Just then the building inspector happened along. His verdict: "You'll need to install a sump pump in the basement." We didn't really like that idea but something had to be done. It was a hot afternoon, so Peter and I decided to do some work in the basement during the heat of the day. We knew the house was built on a solid flat piece of rock but there was a huge pile of sludge covering it that needed to be cleaned out. As I paused to lean on my shovel near the chimney area, my shovel started to sink in the sludge. I remarked to Peter, "I think there is a cavity in the rock in this spot." I began to dig and probe and before long I was down the length of a long handled shovel uncovering a hole about three feet long and a foot wide in the shape of an eye. Our neighbour had allowed us to hook up a hose for any water we needed so I filled a pail and dumped it in the hole: it disappeared quickly. A second and third pail did the same. We discovered a crevice from this hole toward the back corner of the house. Amazingly, there was an old 4 inch pipe cemented in the old foundation leading outside to the footing. Here was a natural drain that would handle all the water. Later, after pouring new cement on the basement floor, we boxed in the hole for a sump pump but have never yet had to put one in. Thank you, Jesus.

We had acquired a large metal front door complete with glass side panel at an excellent price because it was smoke damaged – a bit of sanding and a coat of paint and it looked as good as new. The house in Toronto from which we had secured a number of doors included an outside door complete with screen: the $20.00 price tag did not include a key. Since we now had quite a bit of material and tools in the house, Peter suggested that we remove the lock mechanism from this side door to get a key cut for it which I did. Piling tool boxes and heavy materials against this door for the night, we exited from the front door. As I locked this door, Peter exclaimed, "Wouldn't it be interesting if that key fit the lock in your hand. " I tried it and sure enough it was the same make: one door was new, the other about 20 years old. We only needed to have the tumblers adjusted and we could now lock both doors!

We decided to double insulate the house since we planned initially to heat with a wood stove to save the expense of putting in an oil tank and burner since natural gas had not yet been installed in our area. Our brother-in-law, Jim Sheardown, enabled us to buy an airtight wood stove at an excellent price. We had the baseboard electric heaters for back up heat if needed. One Saturday morning, just when it was needed, a team of men from Mountain Bible Church (now Hill Park Bible Church) arrived. Some ladies from the church brought in food so the men could work through the day. In one day all the insulation was installed.

We still needed to drywall the entire house. That summer the federal government had a building incentive which offered funds for renovations. We applied for the maximum amount of $3,000, a lot in those days and, praise the Lord, it was granted. As mentioned earlier, we had taken one Saturday off to attend our family reunion. We were at Florrie's brother's cottage for the day and Ron was asking us all about the house, how large it was, how many rooms, what we had accomplished so far, etc. A few days later, in drove a huge boom truck from "Rout Building Supplies" loaded with all the dry wall for the entire house along with all the weeping tile for the septic system we needed to install. What a huge boost this was. Dry walling is heavy work and requires a certain amount of skill. We contacted a two man team and they did the whole house for us. What do you think it cost? $3012.00. The grant from the government along with our $12 covered the entire job.

When we first bought the property, my wife had pictured us living in a tent in the backyard of the house during the summer while we perhaps got a kitchen, bathroom and living area completed.

The Miracle House

Again, the Lord had other plans. Since the parsonage was not needed immediately, we were allowed to remain there until the end of October. On our actual moving date, we were madly nailing down the subflooring in the bedrooms so we could lay out mattresses for the kids to sleep on that night. By October 31st, 1982, we had a five bedroom home ready (almost?) to move into which we named "The Miracle House".

Again God provided as years later, the city bought a portion of land at the back of our property for a road. This enabled us to pay down some of the mortgage. We still live in our Miracle House and God's provision throughout the years has been amazing.

4

A VISION IS BORN

"How then shall they call on Him in whom they have not believed? and how shall they believe in Him of whom they have not heard? and how shall they hear without a preacher?" Romans 10:14

MY MINISTRY WITH GOSPEL RECORDINGS INVOLVED OVERseeing missionaries as well as distributing records, cassette tapes, players and teaching materials. When I joined Gospel Recordings, the office in Toronto. was housed on the second floor of a warehouse. Below was a silk-screening company and late into the afternoon fumes would seep up into the offices. At the other end of the building on the ground floor was a company refurbishing 45 gallon oil drums. Every morning we were greeted with the "bong, bong, bong" of the drums being unloaded. When it rained, we scurried for the buckets to catch the water leaking through the ceiling. Besides all this, the new owner of the building was going to increase our rent substantially. Needless to say, we needed better office space.

In September of 1984, I saw a place in Hamilton on Aberdeen Ave. It was a three storey building with a garage that could easily be renovated to accommodate us since we no longer needed space for stamping out records. The building was being sold for a good price so we figured we would need about $90,000 for the purchase and the needed renovations; at the time we had only $9,000. I sent out a letter to our prayer supporters asking people to pray for wisdom and guidance in this venture. A few days later, I was at a conference in California and received a call from our excited treasurer. "Guess what, Allan, we just got a cheque for $90,000 from someone out west." Praise the Lord! The real estate agent needed a certified cheque for the purchase of the building so we deposited the money into our Toronto account, waited the appropriate number of days for clearance and then proceeded with the certified cheque for our agent drawn from our new bank in Hamilton. Now we needed to complete all the repairs and renovations before the end of the year when our contract would be up on our Toronto office. Everything moved along well and our staff relocated in Hamilton. We were also able to rent out the garage and the top floor of the building. This almost covered the cost of overhead for the building freeing up funds for the ministry. We were rejoicing in the Lord's provision of this new location.

The following Spring, I was in Nairobi, Kenya for a field conference with our African staff and I got another call from our distraught treasurer. "Allan, guess what? The cheque for $90,000 that we used for our new facilities was N.S.F. and the bank wants their money back. I mentioned that it was a certified cheque: "How could this happen?" My first thought was *"The Lord giveth and the Lord taketh away; blessed be the name of the Lord." (Job 1:21)*. As soon as I returned to Hamilton I started checking into the situation. We had spent almost $20,000 of the money on renovations and had all our staff relocate to the city. It would be next to impossible to turn back the clock on this one.

The original money had come from a small community. It had passed through the small community bank, the provincial bank, the clearing house in Winnipeg, our Toronto bank and the bank in Hamilton. We were in consultation with our lawyer throughout this process. Further complicating the issue was the fact that we had issued a charitable donation receipt for the funds. In actual fact, the banks slipped up and didn't recognize the mistake until well over six months beyond the issuing of the certified cheque. They realized they were at fault but asked us for the money anyway. Eventually they came back to us asking for 50% of the money, then 25% and then 10%. Our lawyer said, "Allan, you originally had $9,000 in your building fund, that's 10%. Pay that to the bank with an agreement that they will have no further recourse in the matter". This we did and the building served us well for all the years I was the Canadian Director.

McGuirls 1981 Betty Lynne, Florrie, Loralee, Allan, Jr., Allan, Sr., Shari Lou, Ruth Anne

Valerie Deguchi (now Brewer), our field recordist stationed in Nairobi, and I travelled to a remote village, where we were going to do some recording. Four missionaries had flown in with the two of us. The chief invited us to the long house for supper, which was a great honour. During the meal, the chief carried in a large roasting pan containing a stuffed cow's stomach. The chief cut a piece for each one of us. I looked at it and it looked at me with all this green stuff inside. I ask the missionary nearby what the custom is here. In some places you leave some food on your plate to show you are satisfied. Guess what? It's a "cleaner-upper". I tried a little but it was too tough to chew.

Just then, one of the missionaries at the far end of the table was leaning back drinking hot coconut milk. He lost his balance crashing to the floor. While everyone was distracted, I took that lovely piece of cow's stomach, jammed it into my pocket and replaced it with a spoonful of vegetables and carried on. Later, on our walk back, I threw the tasty morsel to the dogs whom I'm sure enjoyed the treat.

In the Congo among the Kinindo tribe I slept in a bed I describe as a barn swallow's nest built out of mud and straw. It's an igloo-shaped mud hut made of cow dung and sticks. The "nest" was like a pocket built into the wall about five feet long with some straw, along with its resident insects, thrown on the bottom. Fitting a 5' 9" frame into the space was quite a challenge. One of the nights, I awoke about 2:00 a.m. scratching and trying to get comfortable. I pulled my little penlight flashlight out of my shirt pocket and saw various 6 legged creatures on the walls of the hut. Being rather cramped, I opened my pocket New Testament and in the dim glow of the flashlight, guess what verse appeared. "Philippians 4:4 *"Rejoice in the Lord always, and again I say rejoice."* Thank you Lord, I get the message: be content where you are. I fell back to sleep.

Another time during this trip, I was sitting for supper in a Kenyan home when a huge flying ant came into the small thatched house. As a delicacy, this fresh offering was presented to me as the honoured guest. Not the venturous type when it comes to food as graciously as I could, I declined.

In another African village, I was invited for supper and while the wife was preparing the food her toddler squatted down and wet the floor. In one motion, the wife stooped down to wipe up the puddle, then using the same cloth wiped off the top of an old chair, sprinkled flour over it and patted out some pastry. I prayed there was a good hot fire for baking!

One of the saddest times was when I was up in Lake Turkana in northern Kenya distributing cassettes for the people there. It was during the Ethiopian famine and many refugees had flooded across the border.. Oh, such poverty, the desperate shortage of food and water! The couple of days we were there, we were given only a few inches of water in a basin: that would have to do us for all our needs. We can never grasp the hardship many of these people go through on a daily basis. Dedicated workers with Gospel Recordings labored from village to village to record God's Word into the languages of these isolated groups. Who can tell how many have come to saving faith through their efforts? One day there will be a great celebration!

As I travelled with Gospel Recordings, I realized that there were a lot of Christian radio broadcasts but most people outside of the cities did not have radios or electricity. Battery powered items were not of much use since there were few places to buy batteries and if people had money they needed to use it for more important necessities. If only they had solar-powered radios, they would be able to hear the Gospel message, come to know the Saviour and get the Bible teaching they needed for growth in their Christian life.

In early 1988 I returned home from this trip praying for wisdom and headed to my basement workshop. Soon I had a solar-powered fix-tuned radio. Fix-tuned to a Christian radio station in the area, the recipient would hear the Gospel message for sure. In North America it is difficult for us to appreciate what our world would be like if we could not read. We are so dependent on the printed page. However, non-readers comprise over

75% of the population in many places and in many language groups it often approaches 100%. Many cultures are, by tradition, aural – meaning most information is passed on through the ear gate. How can we best reach these people with the Gospel if not by radio?

Our Canadian Board was eager to move ahead on this, but other Gospel Recordings centres were not so interested. What was God's plan in all of this? I believed He had given me this burden and enabled me to build a prototype. In the weeks that followed, with the prototype shelved, I felt a distinct inner turmoil. What now?

5

GALCOM BEGINNINGS

"... and a threefold cord is not quickly broken." Ecclesiastes 4:12

NOT LONG AFTER THIS, OUT OF THE BLUE I HAD A PHONE call from Tiberias, Israel, from a total stranger, Ken Crowell, an engineer and ordained pastor. Some years earlier God had called Ken and his wife Margie to set up a" tent-making" business as a ministry base, manufacturing stubby antennas for the electronics field. Their fledgling business, Galtronics, was beginning to prosper. Jews and Arabs, believers and unbelievers were working side by side on the same bench. Eventually the Peniel Fellowship was formed for Messianic Jews. It has now surpassed 400 believers.

Around 1988, at the same time God had planted the vision in my heart, God had impressed on Ken's mind the need for a fix-tuned radio. Ken was puzzled since he didn't work with radio at all. However, when the inner promptings continued, he had his engineers draw up a design. He didn't know what to do with it so shoved it in the bottom drawer of his filing cabinet.

Some months later, Ken was in Florida for a Tentmakers Conference and found himself sitting across the table from Harold Kent, a Florida businessman. When Harold discovered that Ken was an engineer he asked, "What do you know about making fix-tuned radios?"

Ken was astounded and shared with Harold how God had told him to design a fix-tuned radio. "I did just that, but I haven't known what to do with it; it's lying at the bottom of my filing cabinet."

Harold in turn was astonished and related how God had impressed on him the need to flood the world with radios – fix-tuned radios - to share the light of the Gospel. In his mind, he could see a huge globe with lights beginning to blink on all over the world.

Ken added one more thing: he had just read a missionary article about some missionary in Canada who had made a solar-powered fix-tuned radio. Excitement mounted in the two men, so that when Ken returned to Tiberias he pulled out the missionary article and telephoned me. That was the unexpected phone call.

In November 1988, Ken flew to Hamilton to meet with Florrie and me and checked out the prototype I had made. After numerous discussions with him and with Harold and Jo Ann Kent we were convinced that God had brought us together. In February 1989 at the NRB Conference, we three men met together: men from three different countries who had not even known about each other prior to this time. We made a commitment to partner together as the Lord directed to send fix-tuned radios around the world.

Ken returned to Israel and pulled out the radio design file. Since I was still with Gospel Recordings and felt it would be a conflict of interest to actively pursue any missions, missionaries or Christian radio personnel regarding the need of fix-tuned radios, I provided Ken with numerous contacts. He sent letters out to many of these to see what interest there was in this type of radio but it was not long before we realized mail sent to Ken or Galtronics was being intercepted, opened, read and resealed.

Much of it never reached Ken. It was apparent that certain interest groups were not happy with any "Christian" endeavours being conducted in Israel.

It was agreed that Ken would make the radios in his factory in Tiberias, Harold would fund the venture and I, with the knowledge of missions, missionaries and global needs, would find out where the radios were most needed and facilitate getting them there and distributed.

This, then, was the beginning of GALCOM. God brought together three different couples from three different walks of life, from three different countries to birth this ministry. To God be all the glory!

Founders Jo Ann & Harold Kent, Allan & Florrie McGuirl, Margie & Ken Crowell

Florrie and I resigned from Gospel Recordings and in early August 1989 left for Tiberias. There we met with Ken Crowell and Noah Garaway, who was in charge of production for Galcom Limited along with his wife Gila who did much of the sourcing of parts.

We did a lot of detailed planning for the ministry during those few short days in Israel and forged our mission statement:

"Providing durable technical equipment for communicating the Gospel worldwide".

We set out goals, principles and practices. We would partner with other evangelical organizations from around the world to share the Gospel through technology. We would operate on faith, paying as we went. We would bathe the ministry in prayer. We drafted a doctrinal statement and discussed how we would work things out together.

Our new name would be GALCOM INTERNATIONL. "Gal" means wave or commitment (like Galilee). "Com" is for communication. "International stands for worldwide. So GALCOM is committed to communicating the Gospel to the peoples of the world – particularly in their own language.

The three founders – Ken, Harold and I – agreed that GALCOM's main office should be established in Canada, in our home initially. After 10 days in Tiberias, we returned and on August 15, 1989, we pushed aside our dining room table to make room for a desk, our computer and a fax machine which Harold had supplied. (FAX was the best way to communicate across the miles in those days!), set up a workshop in our basement, and at 9:00 a.m. committed this new ministry – GALCOM INTERNATIONAL - to God in prayer.

Florrie and I went to our supporters to explain what was transpiring and how God was calling us into this new revolutionary ministry of providing little solar-powered radios for the world, pre-tuned to Christian radio stations where people in their own language would hear the gospel message. All but one supporter caught the vision (that person soon changed his mind) and graciously continued our support, even though as yet we had no official name or by-laws so we had no way of issuing charitable receipts. Again we had confirmation of God's leading. For this new venture we knew we would have to rely on the Lord more than ever to guide and help us.

Well, now! Where and how does one begin to tackle such a tremendous task? Day after day, while Florrie looked after the administrative duties of finances, correspondence and publications, I contacted ministry after ministry putting out feelers and looking for opportunities to place fix-tuned radios in the most needy areas of the world. Every ministry, whether national or international, would have to be able to sustain whatever Christian broadcasting was available within the country. Frequencies would have to be reliable, a good track record for the ministry was necessary and faithfulness in teaching God's Word with integrity became primary considerations. God began opening doors. Some of the first radios, after the initial 40,000 went into Lebanon and the Middle East, went among the blind in the Philippines, into the jungles of Zaire and Peru, the mountains of Mexico and the northeast regions of Russia.

We needed to develop some promotional materials and I built a large portable electronic map showing the areas of the world unreached by the Gospel, areas we were targeting to send in our little Go-Ye radios.

Soon we needed volunteers to help with literature and other areas – from the beginning, volunteers have played a huge role in the ministry. Without their prayers and financial support, the amazing hours they've given – without all their help GALCOM could not have developed the way it has.

Our first volunteer was Kathy Ward who arrived in October 1989.

Then the first Sunday that I shared the ministry of GALCOM in Paramount Alliance Church, I mentioned our need for office volunteers because we had no funds to pay anyone. Doreen Gibbons approached me after the service and soon was faithfully serving the Lord. She's still with us today, working on the assembly line with her husband Norm who joined us in 2003.

Cathy Ward & Doreen Gibbons our first volunteers

Another couple came on stream in the spring of 1991, when I spoke at Winona Gospel Church, Roelof and Shirley Poot. We really needed representatives to help at mission conferences but also needed a handy man at the mission as we were moving the office and workshop from our home into larger facilities. Roelof was a tremendous worker and helped build special equipment such as video screens we used at the time for showing the "Jesus" film in many countries. Shirley helped with literature, office work and looking after bank deposits – the funds were beginning to come in for which we heartily thanked God.

Then in 1996 the Poots moved to British Columbia. For fourteen years, they capably represented Galcom at Mission Fest Vancouver and Edmonton, as well as at many churches and conferences in Western Canada. Requests for GALCOM radios were pouring in from all over the world and our little team was kept extremely busy.

There are so many volunteers locally that have given so much in the Hamilton area both in labour, prayer, financial support and encouragement. Many have assisted with the mailings, worked alongside our staff in radio production, served on the board and assisted with building maintenance. What a blessing. I know that GALCOM would not be what it is today if it had not been for the volunteers who have literally built tens of thousands of radios right here in Hamilton. Thank you, Lord, for all that these dear servants have done.

Shirley & Roelof Poot served as representatives in Western Canada

So many other volunteers have been donors. Besides the huge contribution of Harold Kent, one of the founders, there have been countless churches, individuals, seniors groups, children's groups, mission agencies and conferences who have sponsored radios, radios stations, trips, audio Bibles. The list goes on.

Other volunteers are not so visible; they are the prayer warriors. Day after day they commit the needs of the ministry to the Lord, pray for those who work on the home front as well as those who travel abroad. They sincerely seek God's guidance, protection and wisdom for all that takes place at GALCOM. They regularly pray for our families. What a blessing to know that our heavenly Father sees all of these volunteers and as the great Judge will not forget one of them. He has His own way of rewarding His servants.

6

RADIO KAHUZI WITH RICHARD AND KATHY MCDONALD

"Thou shalt not be afraid for the terror by night; nor for the arrow that flies by day." Psalm 91:5

WE AT GALCOM HAVE BEEN SO PRIVILEGED TO WORK WITH Richard and Kathy McDonald and Radio Kahuzi in D.R. Congo. This faithful couple has maintained Christian broadcasts in that area for over 25 years through war, famine and disease. Many times their lives have been threatened, but God has protected them and enabled them to continue broadcasting even as bullets ripped through their home. For 17 years they remained in the country without any break in order to keep the broadcasts going: they knew if they left, they would not be allowed back in. Electrical power was sporadic at best, there was no e-mail or internet of any account, no friends to visit, not even the opportunity to see a dentist or doctor.

Our relationship with the McDonalds began in 1994 when hundreds of thousands of Rwandan refugees, fleeing the genocide in their own country, flooded into the 30 camps close to the McDonald's radio station. Richard and Kathy were daily distressed by the desperate plight of the shivering families, widows and orphans huddling together in the chilly nights of the rainy season to find some warmth. Any bit of plastic held over their heads by a stick gave minimal protection.

The McDonalds sent out a heart-wrenching call for blankets and Galcom radios. In response, thousands and thousands of radios wrapped in baby blankets were packed into plastic bags, along with a piece of candy and a Gospel of John in the Kinyarwandan language. These were quickly distributed among the people. It seemed so little among the millions of needy refugees.

Since the need for the solar-powered fix-tuned radios was far greater than the supply, the McDonalds established Radio Clubs in every direction around the station as far away as Bujumbura, Burundi. Each Club had to have at least 50 signed members to get one Galcom radio. Some clubs have more than 800 members. Club members correspond with the station regularly and center their personal lives around Christian service in their villages. They use the station to share their faith by sending letters that are read during the half-hour correspondence program twice each day. They can also request Christian songs to be directed to their families and friends.

God has blessed the ministry and outreach of each Radio Club in different ways. Several Clubs joined together to reach and win the hearts of street kids, widows and orphans. One of these Clubs has become so successful that the new government invited some of the members to the capital city Kinshasa to work with the street kids there. God blessed another Radio Club which began to work with pygmies in the jungle teaching them to read and write while sharing the Good News of Jesus Christ with them. Now the pygmies have their own Radio Club whose president, one of the first educated pygmies, announced his willingness to teach his own people. They are writing letters to Radio Kahuzi and signing their own names. Since listening to Christian radio, Pygmies are seeing the importance of Christian marriage and want to have a Christian marriage for themselves. This is a first for their tribe.

Other Clubs practice agriculture, fish farming, animal raising, beekeeping, tree planting, street cleaning, improving sanitation, repairing roads and bridges and cleaning out ditches. These efforts have significantly reduced the disastrous effect of cholera and other diseases. Other Clubs take up offerings to help members in distress. One girls club is training its members to become better future homemakers and Christian mothers.

Priority for radio distribution has always been to displaced persons, hospitals, prisons, etc. One man was in prison for beating his wife and pouring hot oil on her in a fit of rage. In prison he began to listen to a little portable Galcom radio he'd been given, and while listening to Dr. James Dobson's "Focus On The Family", he was convicted in his heart. He was reconciled to his wife and family and is now an evangelist.

Gila Garaway, the wife of Noah Garaway who organized the first production of Galcom radios in Israel, visited Bukavu and was amazed to see Galcom radios everywhere! The man searching her bags at the airport was wearing a Galcom radio tuned to Radio Kahuzi. Officers in government buildings were listening to Galcom radios. Shops in the big markets had radios playing at their loudest volume. God is using radio in amazing ways to bring the message of salvation to all who will listen.

Kathy McDonald told this story: "One day after church, a lady came up to me, all smiles. 'Do you remember me?' I had to admit I did not. She laughed again and told me she used to sit in our Sunday School class as a little kid years ago. Now she was a grown up Mom. I gave her a little Galcom radio. A few months later, I met this same lady again and she told me that she'd just gotten out of prison and was so thankful she had the little radio to listen to while she sat in that place. She hadn't done anything wrong but her son had; the police couldn't find him so locked her up instead! They finally found the son and put him in jail. He is out now, but his Mama sure appreciated having the radio to encourage her in those hard days of jail time and to share the programming with those who listened alongside her. "

In many instances, the little red (short wave) radios have saved people's lives in the war zones. One man was on his way home early in the evening when a soldier stopped him and demanded all his money. This was a serious problem because the man didn't have any money, but he did have his Galcom radio. The soldier, satisfied, took the radio which saved that man from getting beaten up or worse! Now they're praying that this soldier will listen to Radio Kahuzi and accept the Lord as he hears the Word of God in his own language.

7

RADIO OUTREACH – GUATEMALA

"How beautiful upon the mountains are the feet of him that bringeth good tidings, that publisheth peace; that bringeth good tidings of good, that publisheth salvation." Isaiah 52:7

IN THE OVER 50 COUNTRIES I'VE VISITED, GUATEMALA IS ONE of the most delightful. A very special couple there are Fausto and Miriam Cebeira, who now have five radio stations in remote areas of the country. One memorable visit we went to distribute radios up among the Ixil (pronounced Ee' sheel) Indians in the northwest part of the country. I took with me retired business man, George Calder, who'd always wanted to go with me on a trip.

George and I flew into Guatemala City where we were met by Fausto and his driver, and headed to their home. I should mention that Fausto has a double role: he's a very successful businessman, and an ordained pastor. I'm sure a lot of his earnings have gone into setting up the radio stations around the unreached areas of Guatemala. Next morning, George, Miriam, Fausto and I start out with a load of radios in his four wheel drive vehicle for a five day trip. I remember it well; it had been wet and somewhat cool weather.

We head up into a village where some radios have already been placed before and we've brought more to distribute now. Travelling along this old mud and gravel road, we come parallel to a roaring river. Up ahead, a small tree has fallen across the road. We stop, drag it away, then drive on for a short distance. We notice ahead that half of the road has been washed away, literally. Ten feet below, turbulent water is churning its way downstream. Now what do we do? Go on, hoping the road will hold us or turn around? Prayer is essential and we sense that we should go on. Very slowly, we edge ahead and safely reach the other side. I look at the flow of the muddy water below: if we had ever fallen into that ten foot drop, who knows where we would have ended up? Thank you Lord for watching over us.

Some distance ahead we notice women coming back from working in the fields. Miriam says, "Here's a terrific opportunity to give some radios out to families who don't have one. I remember as if it were yesterday, I approach one lady and with Miriam's help as interpreter ask if she would like a radio. She is so delighted and holds out her hand, a baby on her back. The sparkle in her eye tells the story as she turns on the radio and is able to hear a message in her Indian language. George is thrilled to give a radio to a family he knows will soon hear the Gospel for the first time... then to another family and another. To see the response of these people, so poor, who live in the hills with so little, is gratifying.

We end up eventually at our first destination and the Pastor of the village soon rounds up the people. We hold an impromptu meeting for over a hundred people. Pastor Fausto preaches and then asks for a show of hands of people whose lives have been touched by the radios given out previously. About half the crowd raise their hands. Then he carefully checks on families that still need a radio and makes sure they don't give out duplicates. Miriam introduces me to a lady whose life has been radically changed from the depths of sin to become a glowing Christian. Wow, it is worth it all just to meet this lady. How many people will we meet in eternity because of these little solar-powered fix-tuned radios? What a great work God is doing through these "mini missionaries"!

Miriam then hurries us up as we have to get to our mission base among the Ixil Indians before it gets too dark. As we start out the rain comes down in torrents. Fausto is struggling to keep the vehicle on the road. A short distance out of town, in the pitch dark with the rain still pouring down, he's trying to keep in the tracks of the road when all of a sudden we slide sideways into the ditch.

Stranded in the muck in Guatemala

Now there's a real concern because they tell us there are bandits in the area and it's not safe to remain there since they love to catch foreigners to hold for ransom. So, leaving Fausto at the wheel we all get out to push in the pouring rain. Man, we are not going anywhere! I still remember standing in the middle of the road in the drenching rain and praying, "Lord we desperately need your help." Within minutes, we hear a lot of shouting and noise. Miriam, I see, is trying to discern if it's good or bad. To our amazement, the villagers somehow had sensed our need and came running across the field down the hill to where we were. To watch them work is amazing. With one small light to see, they get the jack out and put it under the left front wheel and then as they raise the car they pile stones underneath. They do the same on the back wheel, then repeat the process on the remaining two wheels, even lying down in the wet mud to make a bed of rocks as a path to drive out. In about twenty minutes we're on our way, soaking wet but rejoicing. We continue on our journey praising God for another answered prayer.

In about an hour and a half we arrive at the Ixil Mission Base where Fausto has a number of workers operating the radio station. Is it ever good to hit the bed that night after cleaning up. The next day we check out the radio station. I look over all the transmitter settings and everything appears to be working well. We meet his radio staff and again give out radios distributing them so that each home gets one.

That night at supper the Ixil give a lovely handmade travel bag to both George and me. And then they give me an actual Ixil hat with long curly ringlets, like dreadlocks, all around the back outer edge.

This community of several thousand has been under the power of a witch doctor. A few weeks after the witch doctor gets one of the Galcom radios, the people find him outside burning all his fetishes connected with witchcraft. He's been marvelously saved and his testimony causes many others to come to Christ. Once leading his people in witchcraft, now he is leading people to the Lord.

This Ixil lady will hear God's Word daily with her new Galcom radio

A funny incident happened at the compound where we were staying. Fausto and Miriam had a maid who did all the cleaning of dishes and helped with meals. Apparently, at night you don't go out wandering around even to the bathroom. As George and I are finishing breakfast, this little maid comes to Miriam and says "Where do I put this thing?" It's a little potty used at night time; she'd just washed it with the dishes which we were presently eating off. Miriam burst into laughter but George's face registered shock. I said, "George, you can't do a thing about it now, it's all in your stomach." We, too, had a good laugh. The trip back to Guatemala City went much better and on the trip back home George and I were reminiscing about the great opportunity we had to assist in sharing the Gospel through these little portable radios.

On the fifth anniversary of Radio Ixil, Miriam reported on the great impact radio ministry is having in this area:

"The three day music festival is in progress with Gospel preaching included – all under a huge canopy on the street corner just outside the studios of Radio Ixil. The four converging streets have been shut to traffic. In the centre square there's room for a platform, sound equipment, musicians and their instruments, and benches to seat approximately 600 visitors. The entire countryside is resounding with the music and God's word carried by the sound waves and captured on tiny Galcom radios. The fact that only the Ixil language is being used has drawn tremendous response. One listener expressed it this way, 'We are especially grateful for the Gospel message in Ixil. Spanish is hard for us to understand. But preaching in Ixil is wonderful!'

"People have come from far and near to the celebration. Some have biked muddy paths, scaled high mountains and slipped their way over rocky trails, crossing swift rivers during two or three days to attend the festival. All are dressed in their festive best. The women's outfits are especially colourful: a large embroidered headpiece, intricately handwoven blouse and belt, bright red skirts and many colourful shawls that secure a sweet baby to their backs.

"The joyful music is praise and worship to our eternal God, overflowing from the hearts of the redeemed. We rejoice to have reached these wonderful people in spite of nearly forbidding barriers of culture, language and great walking distances to their villages. The radio waves swiftly transcend the deep ravines, steep mountains, deep valleys and treacherous rivers. And Galcom radios take the Gospel to the poorest and the most needy. How almost impossible it would have been to reach the thousands of little huts scattered over nearly impassable terrain! One elderly man lamented, 'I am blind. I can't walk very fast or very far. I can't get to church.' But the Christians of Radio Ixil took him a Galcom solar-powered fix-tuned radio. Now he has church at home all day, every day. With tears flowing, another blind man rejoiced, 'I have heard the Gospel in Ixil, my own tongue. I understand it. Now I can live the Gospel.'

"A 100 year old woman in the village of Xebitz was nearing life's end without hope for eternity. Her son was anxiously praying and searching for an opportunity for his mother to hear the Gospel. The Lord answered through the daily program for women. The son conveniently turned the volume up on the Galcom radio and gently invited her to listen to 'good advice offered by another Ixil woman in our own language.' Soon she was saying, 'I want to receive Jesus'. Then she wanted to be baptized. She became a rejoicing Christian. But her conversion was none too soon; shortly afterwards she became ill and slipped away into Jesus' arms in heaven. Thank God for Galcom radios!"

Miriam continues, "As far as we can see there's only one problem with the Galcom radios: WE HAVE TOO FEW! We have given out a limited number in selected villages, but many people are desperately pleading for 'just one more for me'. To cover the area we should have 10,000 more radios.

"One Christian was desperate for a Galcom radio. Max, the mission director replied, 'Your village is not yet on our program'. The man begged. Max said no. The man insisted that just one wouldn't hurt. Max said, 'No way'. But the man finally wore Max down with his entreaties and Max gave him the only radio to be had in this remote village. The man was overjoyed. He said, 'Brother Max, I will be returning home now. If I never see you again here on earth, (very emphatically) WE WILL MEET IN HEAVEN!' Max was impressed with the way he made his statement. The two men parted. The Christian took the tiny path that would eventually lead to his village. He must cross mountains and rivers and trudge for many hours before arriving. Within a few hours, the news came back to camp: the brother with the radio had drowned! His feet had slipped from the crudely made barge that was taking him across a swift river and he was swept downstream... straight into heaven!"

Praise God for what He is doing not only in Guatemala but all over the world.

8

PRODUCTION MOVES TO CANADA

"As for God, his way is perfect: the word of the Lord is tried: he is a buckler to all those that trust in him." Psalm 18:30

MAKING THE TRANSITION

In late 1994, with over 160,000 Go-Ye radios already distributed around the world, the burgeoning Galtronics business in Tiberias required more production space: the founders decided to move radio production to Canada. We sent our new electronics designer, Tom Kerber, to Galtronics to learn what was involved in radio production. The most extensive, heavy fog in airport history, kept Tom grounded in Toronto for two days. Thankfully, he obtained the last seat on a Saturday flight and finally arrived in Tiberias.

The Galtronics engineer was giving Tom a tour of the plant. As Tom followed behind the engineer, suddenly a long, heavy, four-inch diameter, steel pipe fell between them crashing to the floor. The construction worker installing a sprinkler system overhead was so apologetic: he said that while he and his partner were raising the pipe into place, it just became so heavy they could not support it any longer. With an object of such weight and force, Tom could have been seriously injured, or worse. Tom was just praising the Lord for watching over him! Amazingly, within a week Tom was able to grasp the entire production process.

During the following months, equipment was shipped to Canada and soon we were in full swing. In March, 1995 the first Go-Ye radios rolled off the Hamilton assembly line with a redesigned circuit board, thanks to Tom. In the 17 years since, we've produced over 800,000 more solar-powered, fix-tuned radios in Hamilton, Praise the Lord!

Another interesting development was the provision of facilities in Hamilton to accommodate the radio production move. As mentioned earlier, we started the ministry in 1989 in our dining room, kitchen and basement. Within a year we'd outgrown our home and were able to make comfortable arrangements with the Mountain Bible Church (now Hill Park Bible Church), the church I'd been pastoring. Up to this point, Florrie and I handled all the orders, conferences, publications, bookkeeping and secretarial work. We now needed a full time secretary. The church agreed that GALCOM could use one of their large rooms for Galcom rent free if we could provide them with several hours per week of secretarial assistance. We hired Debbie Ruser as our first secretary and this arrangement worked beautifully for both parties for several years.

However, by Fall of 1994 we were outgrowing this church space. I looked around for something larger, praying that the Lord would guide us. I found 4,000 square feet of space in an industrial mall nearby. At $500 per month with a five year lease - a $30,000 commitment, I agonized over this contract for some time. In the end, I believed this was where God wanted us and in November proceeded to sign the papers: we were to move in on December 15th. It was during those few short weeks between

November and December as we were preparing to move to our new location that we three founders made the decision to move radio production from Israel to Canada. God already knew we would be needing the additional space! Within those few short weeks, we'd settled into our new facilities, hired production staff and were shipping out the first radios by March, 1995.

Florrie, Shari Lou, Betty Lynne & Loralee singing at the dedication of 65 Nebo Rd.

Radio production continued to develop in Canada for almost two years but in late 1996 and early 1997 funding began to dry up and we had to lay off all of our paid production staff. Overnight, we were cut to a staff of four: Ron Slade, our faithful bookkeeper who was seconded to us from Bible Christian Union, Neil Woods, volunteering in the technological area of production plus Florrie and myself. What was God telling us? Again, we spent much time in prayer asking the Lord for His direction. As quickly as donations had dried up, they flourished again. We hired back Kristina Gomez for radio assembly, hired Neil Woods full time and had Florrie oversee production.

Production shot up to 2,000 radios a month and continued to grow. Early in 2014 we're anticipating the production of our one millionth radio!

To chop the antenna wires, Allan, Jr. built "Charlie" from scraps

115 NEBO ROAD

In 1999, as we approached the end of our five year lease, we realized that the rent would be increased significantly. On top of that, the space next to us had just been rented to a furniture manufacturer. Often, especially by the close of the work day, noxious fumes would seep into our area. We sensed the Lord nudging us to look for other facilities. Under the leadership of our Board Chairman, Ken Hajas, we eventually bought a building just a few hundred feet down the road. Through donations and the money we had already been setting aside for a building we were able to make the purchase. Ken also laid out an ambitious plan for paying off the building in five years by selling bonds, renting some of the space and praying for donors to support the project. God honoured those plans and within five years we were debt free.

The new facility was mostly warehouse with high ceilings. We could envision dividing this area into two storeys for offices and manufacturing area. Just at that time, God brought another man onto the Galcom scene, Don Jeffries. He very capably volunteered as project manager and charted us through this entire process. Later, his wife Gladys served for a number of years in production as well as in our office. Many volunteers gave innumerable hours of work to bring about this transformation and have it ready in time for us to move. We knew that there would be far more space than we would require and the plan was to rent out about half. Tenants were secured which helped to offset the costs of carrying the building. Once again, we could see clearly that God was at work.

TECHNOLOGICAL DEVELOPMENTS

A ministry such as this involves countless hours of technical research and development. In the early years, the technology for the radios was undertaken in Israel. As production moved to Canada, Tom Kerber worked on a number of revisions of the analog circuit boards which powered the fix-tuned radios. However, as this technology aged and the global supply of the needed parts and components became depleted we knew that tremendous changes would be necessary. For some years, several broadcasting ministries had talked about a digitally fix-tuned radio. Prototypes were made in the US, China and Israel along with one designed by Tom but none of these were economically feasible, nor did they match the quality of the long-standing analog models.

In the late 1990's our son, Allan Jr. had just returned from working in western Canada and when he saw Galcom's need for a new design he took up the challenge. Within a few weeks, the Lord enabled him to develop exactly what was needed: a digitally fix-tuned radio that could stand up to the elements in the remote locations where they would be used and within a price range that equaled our present radios. It was soon in production. Shortly afterward, the Galcom Board requested that Allan be hired to do our research and development since Tom Kerber had set out to establish his own company and we had been without anyone in that position for several years. Allan Jr. has been with us ever since and God has gifted him in providing Galcom with so many of its technological needs.

We needed a better method for gluing the radio cases together and Allan Jr. began investigating the possibilities. He felt that a sonic welder would provide a cleaner, faster and more cost-efficient solution. His searching led him to a supplier in Rochester, New York where he was able to purchase a used one at a fraction of its original cost. We are still using that welder to this day.

When Allan Jr. travelled to Rochester to pick up the sonic welder, he also noticed several types of robotic arms. For many months we had talked about automating some of the processes in the production of the radios but robotic arms are extremely expensive. Again we asked the Lord to give us direction. We had some limited funds for this project but needed much more. As I shared what we were doing with several people and churches, they were excited to help with the funding. Miraculously, Allan, Jr. was able to secure five arms, worth over $30,000 apiece, for a total of $5,000. One gentleman who happened to be at Galcom when the arms arrived was so thrilled he donated $3,000 on the spot towards their cost.

Another friend and former Board Member of Galcom, Marcus Verbrugge, worked on commercial conveyor belts for manufacturing. He was able to provide us with a used system which Allan was able to incorporate into the production assembly. When the installation was finished and paid for, a donor wrote a cheque to reimburse the cost. Again God provided.

We also had a need for an inexpensive, versatile, low-powered transmitter. After many months of work, Allan, Jr. was able to develop an incredibly small unit that could also act as the mixer for a radio studio. He needed to conduct extensive tests to have it FCC and IC certified but the Lord led him through the entire process. Today this Cornerstone Transmitter is used in many of our mother stations and repeater sites.

In the mid 1990's, Ken Crowell in Israel had been working with Tom Treseder on an audio Bible. By 1996 they were able to squeeze almost three minutes of reasonable voice quality onto a chip – the absolute limit. We may smile at that today, but it was quite a breakthrough then. This project, called MegaVoice, stalled for a number of years as technology evolved drastically. By 2005 a working audio Bible was available. For the developed world, this was nothing new: the Bible was available on tapes, CDs, MP3 players, etc. However, a unit that contained the entire Bible, needed no external power or player and utilized a speaker as well as a headphone jack was exactly what was needed in the emerging countries. This little solar-powered audio Bible player could easily fit into a pocket and many thousands of them began to be distributed out of Israel, Galcom USA and Galcom Canada.

By 2008, Allan Jr. had been able to incorporate the audio Bible into the radio. The advancements in electronic technology and microchips were so rapid that the industry always seemed to be on a steep learning curve. After a number of revisions, this unit is in production and being sent all over the world. Now people can not only hear the radio messages but can compare it to God's word in their own language for themselves.

It is interesting to note that when the Pharisees wanted Jesus to prohibit his followers from praising Him as the Messiah, that Jesus replied, "If they keep quiet, the stones will cry out." (Luke 19:40) Well, silicon chips are made out of ground up stones, and today they are "crying out" the Good News of Jesus Christ around the world day and night.

From left: Go-Ye Radio, Cornerstone Transmitter, Envoy (Audio Bible) and ImpaX (radio with Audio Bible)

9

HAITI – FIRST LOW POWERED RADIO STATION

"He that goeth forth and weepeth, bearing precious seed, shall doubtless come again with rejoicing, bringing his sheaves with him." Psalm 126:6

THERE ARE MANY STORIES OF HOW GOD SHOWED HIS HAND in guiding us to set up low powered Christian radio stations and to distribute fix-tuned radios in countries around the world. When I speak of low powered radio stations, I'm referring to AM or FM radio stations that are no more than 2,000 Watts in power. Most are 50 Watts and are for small, unreached communities where the Lord has allowed us to bless many of these people. Many stations are 250 – 300 Watts which will reach a good sized town. We have been involved with the setup of over 160 Christian radio stations around the world. Where Christian radio broadcasting was already available, we were able to supply solar-powered fix-tuned radios to be distributed in the area. However, many language groups had no Christian broadcasting at all and were unable to afford high powered radio broadcasting systems or be able to sustain them. With low powered radio stations that operated from a regular electrical wall outlet, the possibility for Christian radio was opened to almost anyone, anywhere. But let me share with you the story of Galcom's first radio station installation.

It was a 50 Watt AM station for Port Au Prince in Haiti in June 1993. Just a few months prior to this project, I was at a Christian technical conference in Colorado Springs. There I'd met a radio engineer, James Cunningham, who told me that over the past couple of years God had shown him how to build a 50 Watt AM transmitter which was exactly what we needed for Port Au Prince, Haiti. Jim had never heard of GALCOM and our solar AM radios so he was very excited to see how we could complement each other in radio ministry. You see, God's ways are not our ways: He brought us together at His right time to set this station up in Haiti, partnering with the Church of the Nazarene and Pastor Simone.

Pastor Simone was working in the slums of Port Au Prince with very little to live on, but a heart burdened for souls. Ross Robbins, of Light of the World Ministries, had been doing a great work in Haiti and heard of Pastor Simone's need. Pastor Simone was running an orphanage on a shoestring and also wanted to set up a radio station so he could minister to the needy families around him. Pastor Simone had a building and some land but no equipment. Ross asked if Galcom could help. Yes, we could. With the recent contact with James Cunningham we could do the whole AM 50 Watt station for only $1,200. U.S. After checking out Pastor Simone's credentials and the viability of the operation we agreed that God was leading us to proceed with this project.

A day or so later, I was asked to speak at a ladies' fellowship in a Presbyterian Church in east Hamilton. Only a small group was there. As I closed the presentation, I mentioned the need for the Haiti radio station and asked the people to pray for this project. After the meeting, we went to the kitchen for tea and a lady spoke to me on the side. She said, "Allan, tomorrow would you drop by my house? I'd like to do something for this station". The next afternoon she presented me with a cheque for the remaining amount of money needed to complete this station. Once again God was providing.

A number of weeks later in June of 1993, I flew to Haiti with a quantity of fix-tuned radios and studio equipment. James had already arrived with his transmitter. God enabled us to go through customs with no difficulty and all the equipment arrived safely. Being our very first installation of a radio station, we made it a practice that we would begin with a session of prayer before starting any work seeking God's wisdom and protection in the midst of the ever-present spiritual warfare in Haiti. I was so glad we did.

The first challenge we faced was space. We were told there was sufficient land to put in a grounding system for an AM radio station. But, when we arrived, we discovered a layer of concrete over the entire compound. Surrounding the area were high walls with shards of glass imbedded in the top to discourage trespassers. Normally, we require a quarter-acre field to set up the ground plane. How could this ever work? Again God answered prayer as near the front of the building, just below the surface we found a metal drain pipe cemented to the city drainage system. It only took a few minutes to clean off the surface, connect the ground strap and there was a perfect grounding system for the station.

Next on the agenda was to install the antenna and guy wires.

We found an old wooden ladder which a couple of fellows helped me position over the glass-studded wall so I could securely attach a guy wire. I climbed the ladder and was just above the wall when suddenly I heard an ominous cracking sound. Eyeing the glass shards beneath me, I didn't care to become minced meat so with a prayer for help, I slid down the ladder and jumped into a pile of sand just as the ladder crumbled in half. Once again I was praising the Lord for His protection.

What a challenge to connect a guy wire to anchor the tower to this pole

With the job not yet completed we hunted around and found a taller aluminum ladder to reach the top of the pole where the guy wire would mount. To reach the height we needed, the ladder was almost in a vertical position against the pole and two Haitian men were to hold the ladder so I could climb up and attach my safety belt and fasten the guy wires to the tower. So up I went again with my harness around the pole to begin working on the anchor for the guy wire. Soon the ladder began to wobble. I looked down and the two Haitians were standing at a distance watching what I was doing. "Get back here you guys and hold the ladder", I yelled as I was 30 feet in the air. This happened several times but eventually everything was secured, the transmitter installed and the studio equipment all connected. Praise the Lord, the station was ready to broadcast.

However, another challenge faced us. It was now Thursday and we had made good progress but a gasoline embargo had been ordered by the USA and Canada. Tension had been rising so much that the American Embassy was closing and all US citizens were leaving immediately. The Canadian Embassy was closing on Friday and they too were advising everyone to leave. What were we to do? Our flights were booked for Saturday and we needed Friday to complete the work. We asked the Lord to guide our steps and decided to stay. By Friday night we were all finished and were safe. We again went up into the mountains to sleep where it was much cooler.

During the night Ross woke up and sensed the Lord telling him to make sure to get on the road in the morning by exactly 6:00 o'clock and not a minute later. So to our surprise, there was Ross waking us up a little after 5:00 a.m. "Come on you guys, we have to be on the road by 6:00." We all got moving and sure enough were on our way in time. As we headed for the airport, what a surprise when we rounded a bend in the road and saw the entire roadway blocked with cars and trucks trying to gas up at a particular station before that gas station ran out of supply.

There was no way we would get through the tangled mess in time for our plane. Just then, a huge army truck came roaring down, cut in front of us and there in the open back we see two bench rows of Haitian soldiers well armed with rifles. They came to a stop directly in front of us and the soldiers jumped out. Ross cautioned James and me not to say anything, just to look straight ahead. As Haitian soldiers, they could easily shoot us and no one would ever know what happened. All Canadians and Americans were hated because of the oil embargo being imposed. They began to look in our windows and as I stared ahead I noticed one large cattle truck blocking the road. A general walked up with his gun, aimed at the driver and I watched as with his left hand one finger goes up: then the second finger. It is clear he is giving the driver the 1, 2, 3 to get out of his way or else. The truck begins to move making a pathway for the army truck. Another car is literally lifted up by the soldiers and pitched off to the slope to make way for them. The soldiers get back into their truck and take off. We breathe a sigh of relief as Ross says to our Haitian driver, "Stay close behind them. Don't leave any gap until we're down the mountain". God again proved His faithfulness and allowed us to reach the airport safely and on time for our flight. Thank you Lord.

That first station that we put in for Pastor Simone was on the air for years. He would start early in the morning spending several hours praying over the radio for different prayer requests that people were dropping off at his station since they had no phones. The church grew to overflowing. Some years later, the Lord called our brother home. I am sure he received a special crown for his faithful sacrificial service over those many years. Thank you Lord for that lady who was prompted to give the money necessary to make this radio station a reality. She too will be rewarded.

10

PENETRATING EURASIAN AIRWAVES WITH THE GOSPEL

"The law of the Lord is perfect, converting the soul: the testimony of the Lord is sure, making wise the simple." Psalm 19:7

MOLDOVA

It has always been a blessing to attend the National Religious Broadcasters (NRB) Conference each year and meet some of God's special people from around the world involved in Christian broadcasting. One such person was Pastor Florin Pindic-Blaj who in 1994 had gone to the NRB conference in February to learn more about radio. Florin and his wife, Lidiana, with Little Samaritan Mission had been involved in setting up a number of orphanages/schools in Romania. As well he had received a license for the first station at Chisineu, Moldova. He was looking for someone who knew something about low-powered broadcasting and in the midst of a crowded session bumped into me. Somehow, I had asked him why he had come to NRB and we soon got talking about radio. He explained that the people there are extremely poor and not everyone has a radio. I said we could help. He was speechless when I told him we could provide solar-powered fix-tuned radios. In June 1994, he got the station up and running.

Shortly after that something amazing happened. We were still making radios in Tiberius, Israel at that time under the leadership of Ken Crowell. Well, they had made 2,000 radios tuned to FM 104.1 for a specific country when, without warning, the government changed their frequency. With the radios already custom tuned to that exact frequency they didn't know where else they could be used. When I heard about the dilemma, I contacted Florin in Moldova and amazingly he had been granted that identical frequency but had no money to pay for the radios.

He got a special arrangement to fly to Tiberius, met with Ken, and was amazed to see the quality of the little radios. Being solar-powered, they were especially valuable since many of the people in Moldova could not afford batteries. He decided to take 200 radios back with him and have the remaining 1800 shipped in. While in Israel, he met some Russian farmers who were shipping a quantity of strawberries to Moldova. They agreed to add the 1,800 radios to their shipment and send them for him thereby covering the cost of shipping.

Flying back home, the president of Moldova, who had also been visiting in Israel, was on the same flight. Pastor Florin happened to know him and when they arrived in Moldova, the president arranged for the 200 radios to be cleared through customs, again at no expense.

So here were radios first destined for one country and God redirected them by a series of unusual events to Moldova. They were fix-tuned to the exact frequency needed. Two hundred arrived with no shipping charges. The remaining 1,800 were shipped for free and entered the country with no tariff assessed.

After twenty years these little missionaries are still proclaiming the Good News of Jesus Christ. Over the years, over 12,000 GALCOM radios have been shipped to Moldova where we can say tens of thousands of people over the years have heard the Gospel with many coming to Christ. How amazing our God is!

One of those who received a radio was Rodica, a nurse in a hospital in Chisinau. She came to know Christ through Little Samaritan Mission (LSM) Radio run by Pastor Florin. After receiving the Lord, she asked LSM for a GALCOM radio to take to work so she could share it with patients. She began giving it to patients for several hours each day and night so each of them could have a turn listening to Christian radio. As she made her rounds, the little radio made its rounds too always bringing the Good News of Jesus Christ. Through this one little radio many patients came to know Christ and she continues this ministry leaving the radio with patient after patient during her shift. The following are some of these stories:

Vladimir was making repairs on his house when flammable liquid spilled and ignited, causing third degree burns over 55% of his body. Rodica would give him the radio at night to listen to the Gospel messages: it was his only source of comfort. Rodica contacted LSM and they came to visit Vladimir and gave him his own radio. He kept it next to his ear night and day. He was so thankful he began to cry. When LSM workers returned four days later, Vladimir had died listening to the Gospel on his little GALCOM radio.

Vasile had an accident at work when a bucket of hot tar fell and burned his body. Like the other patients his desire was to have the Galcom radio during the long nights of suffering. When it was his turn to have the radio he was overjoyed, and even said that maybe God meant for him to be in hospital because now he was able to receive this precious "preacher" of the Gospel.

Anatol was an electrician who was electrocuted by accident. He lost both arms as well as all his hopes and dreams for the future. He said, "My nurse, Rodica, told me that only God could help me through this difficult time of my life. From that moment on I started to listen to LSM radio, and I realized that I was just passing through this life but that I have an immortal soul and must take care of it. I still have many things to learn but I know that this little radio will help me to understand what God's Word teaches.

Zina and Valeriu are invalids. Valeriu is blind and her husband suffered two heart attacks and has liver disease. They said, "Since we became sick we have had to sell everything we had in order to have money to purchase medicine and food. We didn't have a radio but someone told us to go to LSM and they would give us one. I didn't believe it a first, but went anyway and received the wonderful GALCOM radio. Amazingly, since the first day we had the radio our spirits lifted. It seems as though all the programs and sermons on LSM radio are especially for us. I never believed a little radio could change our lives so much in just four months. We have now both received Christ and are so full of joy.

POLAND

About a three hour drive north of the city of Warsaw, Poland is the city of Ostróda with a population of over 60,000 people. A certain lady, Ewa (pronounced "Eva") Brycko, had a burden for this city with only about 20 known believers. Her vision, was for a good FM radio signal that would reach not only Ostróda but the outlying areas to a possible 100,000 people with the saving message of Jesus Christ. She had little money, and attended a small evangelical church in a country not friendly to the Gospel. FM radio was just on the rise in Poland so this was a courageous venture to set up the first evangelical radio station in the country. How would she even begin?

Ewa saw a hundred year old water tower on the highest point in the city which had not been used for years. She knew that the higher the antenna for FM, the farther the signal would reach. She checked out its availability and was able to obtain the water tower for a very small sum of money. After receiving the keys to the building, she was astounded to discover on entering that, lying on the floor was a huge, unexploded bomb from the Second World War. Apparently, it had gone through the roof of the tower, through the floor of the water tank and had fallen clear to the ground floor without exploding.

A WW II live bomb was removed from the floor of the Water Tower

Since it was still live, she had to call in the military to defuse and remove it. This done, she now had the challenge without any money to obtain radio station licensing and equipment.

Ewa had heard of the National Religious Broadcasters Annual Conference in 1996 held in the United States so decided to travel there to learn as much as she could about broadcasting. She raised enough funds to make the flight to Washington and on the second night walked into the huge auditorium around 7:00 p.m. and sat near the back to rest for a few minutes until the evening program would begin at 7:30. Another man, knowledgeable in radio, also needed to rest for a few minutes so came into the auditorium a little early. He sat a few seats from Ewa, and being the friendly sort, greeted Ewa and asked why she was there. She related her story and her need saying she had a license for a radio station in Poland but no knowledge or equipment for setting one up. The man replied, "That's exactly what we do at GALCOM. I'm Allan McGuirl and we set up radio stations around the world." Her eyes lit up! How amazingly the Lord works things out.

She needed money and help to put the station in, and guidance on how to go about it. About six weeks later, I was speaking at The Peoples Church in Toronto and Pastor John Hull asked what project we might have for them. I told him we needed $12,000 for a station in Poland. He took up an offering on the spot of just over $9,000. Praise the Lord! About six weeks after that, I was speaking in Peoples Church, Montreal and the pastor asked me if I had a project. I mentioned that we needed about $3,000 more for a station in Poland. He asked the missions chairman how much money was on hand. Response: about $3,000. He says, "It's yours for Poland." I contacted Ewa and she was delighted. We worked out a date, September 8, for me to go there to work on the station; this would allow a couple of months to secure the equipment required and get it shipped to Ostróda and get the necessary approval. It certainly took all of that time!

I was wondering how I was going to manage all of this when someone generously offered to pay the airfare for my son, Allan, Jr., to go with me. What a blessing. Ewa had also lined up some Polish people to assist with the installation as well. The transmitter was to be installed in the 100 foot tower about 10 storeys high. It was quite a feat climbing the 89 steps around and around from the ground to the floor of the tank above especially when we had to do this numerous times to carry equipment up and down. When we reached the floor of the tank, we needed to be extremely careful to stay clear of the hole the bomb had made as it had passed through to the ground below.

By God's grace the station was all hooked up with the antenna on the roof of the tower. We used a building next to the water tower to set up the studio: one room for "on air" and another for an office. We turned on the power and while everyone in the room held a solar-powered fix-tuned radio, Ewa started talking. The emotion that filled that room was indescribable as they realized they could now share the Gospel clearly and effectively to over 100,000 people in that area. Even the mayor of the town came to visit the station and although he was not of their faith, was rejoicing with them. God works in wonderful ways. A tower that was raised to bring water to the people was now bringing the "Living Water", Jesus Christ.

ESTONIA

Estonia Family Radio broadcasts out of three locations in the country. GALCOM has been able to supply over 4,000 fix-tuned radios to help spread the Gospel there. Pastor Endel Meiusi told of two ladies, too poor to afford their own radio, who were given GALCOM radios. Both accepted the Lord. After preaching in a cathedral in the capital city of Tallin, Endel was mobbed by close to a hundred people asking for the solar-powered fix-tuned radios. In the previous political upheaval most people had lost all of their savings – it seemed everyone was in debt. Endel has had requests from hospitals, prisons, pensioners and a host of families.

Endel approached the warden of one maximum security prison to ask if radios could be given to the prisoners. At first he was not interested but then was willing to take about 250 radios to distribute among them. A number of weeks later, Endel received a call from the warden: He was happy about the radios but had encountered a problem. He first mentioned that the whole atmosphere in the prison had become much more peaceful since the radios had arrived. Many of the prisoners were requesting the Bible studies that were offered on various programs. However, there were over 150 men who wanted to be dipped in water and he didn't know what to do with them. Shortly thereafter, Endel was allowed into the prison to baptize these men who had become believers in Jesus Christ. Several of them were even allowed to take some courses at a nearby Bible Seminary. Nothing like this had every happened before.

Endel could use thousands more radios each year. Programming from this station is provided 24 hours a day and is run jointly by the major evangelical denominations. Even the Prime Minister was listening. When the island station of Saaremaa was off the air for two days due to a broken cable, a newspaper writer had to listen to the regular Estonian station. "It only talked about problems with no solutions. The Christian Radio Station seems to have all the answers," he wrote.

One lady in the city of Tallin was listening to her little radio just after hearing from the doctor that there was no cure for her sick child. Just then, she heard the message that, "God answers prayer and can do the impossible." Right then and there she simply asked God, "Please, heal my child." Miraculously, her child was healed. She then found out that the radio message came from Oleviste Church not too far away. She made her way there and told them what had happened. Shortly after, she accepted Jesus as her personal Saviour and shared her faith with her two sons who were policemen in Tallin. They too turned their lives over to Jesus and attended baptismal classes along with their mother. Once again, a little radio touched so many lives.

MONGOLIA

One of the greatest delights is to meet people from other nations whose lives God has touched in a wonderful way. As they have been obedient to His call, He has led them in amazing ways.

One such man is Batjargal Tuvshintsengel or as he says, "Just call me 'Bat'". I'm glad to oblige. Born in Mongolia during the communist regime, his father was an engineer and a member of the Mongolian Communist Party for over 35 years. His mother was an atheist who later came to Christ in 1999. Bat had never heard anything about Jesus Christ or the Gospel until he was 20 years old. He was studying English and a lady from England started to witness to him. After much thought and investigation, in August 1991, Bat accepted Christ as His Saviour.

Florrie explains to Batjargal how the Galcom radios work

Coming from a communist background, God brought Bat through many circumstances that raised all kinds of questions about God. Through contact with many ministries and especially Far Eastern Broadcasting Corporation, he now has a powerful Christian radio station in the capital city of Ulaanbaqtar, the very first in the country, called WIND FM.

We have had the privilege of sending over 4,000 solar-powered radios to Mongolia where I am sure many have come to Christ. In fact, this station is so popular, GALCOM has just sent a new radio station transmitter to Mongolia to use as a repeater for the city of Bayanhongor with a population of over 30,000. More radios will accompany this station reaching through the open doors in this country. Bat is continuing to reach out to his people with the saving Gospel of Jesus Christ.

ALBANIA

In Pogradec, Albania, George and Nancy Strum with Radio Logos have an exciting radio ministry. The signal even reaches into Macedonia. One Easter Sunday, they were planning an outdoor worship service for their staff followed by a picnic. That very hour George got a call on his mobile phone. The call was from a man in a nearby village who said he had been listening to Radio Logos for many weeks and just that day had opened his heart to receive the Lord Jesus as His Saviour.

Also in Albania, Brother Tani Baraku has an FM radio station in Korce. Ron Marland, one of our volunteer installers, and teams of other volunteers have done a great work in rebuilding this radio station. A few years earlier, Tani's father, Pastor Cimi, who initially ran the radio station in 1991, died suddenly. This left Tani, just 18 years of age, to take over the work along with his mother. Along with his family members and friends they have continued keeping the station growing and they are carrying out a great ministry. Funding is always difficult in these situations but as they trust the Lord, He continues to provide. We were able to supply them initially with 1,000 fix-tuned solar-powered radios to be distributed among the village people, many who have no electricity. We have also partnered with Radio Emanuel located in south eastern Albania. This station covers Korca, Billisht, Erseka, Leskovik and Progradec along with the many villages within the area. Supplying solar-powered fix-tuned radios for distribution among these people is a way to help them reach many who otherwise would not hear. Many listeners visit the station for the sole purpose of getting one of these little missionaries.

Lelezim and Liliana are the parents of four children whose ages at the time were 13, 10, 5 and 3. Lelezim finds the radio so helpful because he can take it anywhere with him. He has shared that the difficulties, temptations and daily struggles are tempered by the faith and hope he receives from the regular broadcasts. Their children are listening to the children's programs every day and learning more about the Lord. The parents see that being raised with God's Word and His wisdom is the best investment in their children's lives.

Another interesting story comes from the village of Bracanj. The pastor of the Evangelical Church in Bilisht took a team of people to distribute some of the GALCOM radios. They met a man working in his fields and started sharing with him about Jesus. He was very open to what they had to say and invited them into his home where he produced a GALCOM radio. He and his family had been listening to the Christian programs for more than a year and he was eager to commit his life to Christ. The Holy Spirit had already been working in his life and that of his family!

During the World Cup 2010 lots of radios were distributed in Erseka. Since there was a high number of young men listening at that time, Radio Emanuel targeted their programming to appeal especially to them. During match break time, the station featured amazing testimonies of soccer players who had accepted the Lord into their lives. The feedback was incredible and of course the little GALCOM radios were much in demand.

A number of fix-tuned radios were allowed to be distributed into the high security prison in Korce. Elvisi, one of the prisoners, began listening to the music especially the song "Praise You In This Storm" (Casting Crowns). As comments were made about the song, he understood the main message: worship God in every situation and in all circumstances. No matter how big the storm is, there is a reason for everything that happens. Elvisi realized he was doing time because of choices he had made. But he was given hope that after the storm things could get better again. He soon started receiving visits from volunteers and enrolled in a Bible Study. He confessed: "When I was free, I abused my freedom; in the midst of my storm, God reached down to me and raised me up again. My life has been transformed and my desire is to continue walking in the Lord."

More recently, I was able to visit Albania to install a new repeater station just east of Korce that transmits not only to the eastern border but also into Macedonia and Greece. We were able at that time to blitz several towns in the area with 3,800 more radios.

God has promised that His Word would not return to Him void but would accomplish what He purposed. Thousands more are hearing the message of God's love and how He changes lives bringing hope and peace for those who put their faith in Him.

INDIA AND NEPAL

It was mid May 1995 when I was asked to visit Bible Christian Movement in India to set up a small transmitter system for simultaneous translation for a conference they were having. This was my main task, but piggy-backing on this I was also to check on the possibility of Christian radio in India and Nepal. I arrived in Hyderabad and met with the BCM leaders and went about installing the translation system. Testing went well and for three days I worked with them to the end of the conference. When we were finished, we left the equipment with them for further use.

I departed for Bangalore and met with Far East Broadcasting Company to see about radio outreach. Bangalore is an electronics centre and so I met with a company to see the feasibility of building radios there if the need arose.

I had an invitation to visit Nepal on my way home. I flew into Katmandu, Nepal from Delhi, India. Pastor Pandry and a friend met me at the airport around 8:30 p.m. at night. I remember driving from the airport on a dark narrow road. Shortly after starting out, we see a whole lot of excitement up ahead. Apparently a fellow on a motorcycle had been struck by a car. As we get closer to the scene, people start pounding on our car window. "Ambulance, ambulance!' Pastor Pandry says. "The policy here is that we have to become an ambulance and take this man to the hospital. Allan, come sit in the front between us". I squeeze in between the pastor and his friend. I no sooner settle on the "hump" than the injured man is dragged by his shoulders and legs into the open car door and deposited across the back seat. The door is slammed shut and they shout, "Go, go"!

Pastor Pandry asks me to try to reach around and comfort the poor fellow who is in real agony. I am squashed between the two front seats and the injured man is hitting me on the back and screaming in pain. Soon I am feeling the pain, too. I reach back to try to calm him down but can't really communicate in his language. Praise the Lord we only have a short distance to travel. They reach the hospital and blow on the horn. Out come two fellows with a big pan shaped like a saucer with handles on each side. They rest one handle on the door frame and drag the poor fellow onto the pan and carry him off. I move into the back seat again not realizing that there is blood splattered all over the seat in several places. There's blood all over the back of my shirt as well where I was being pounded. By the time I get to the guest house, I look like I've been fighting with a cat. That was my welcome to Katmandu, Nepal.

Meeting the next day with Pastor Pandry and several other mission leaders, we investigate the need for low powered Christian radio in the country. We are told that in the future it will be coming and, yes, there is a great need to plan now so that when the door opens we will be ready.

The next day, I had arranged to go to a remote village about three hours drive up into the mountains. Pastor Pandry sent one of his helpers, as a translator, to accompany me in a taxi. We arrived by noon and met with leaders who were interested in a radio station. I did a study of the area to see what was needed for a station and found out details about this group's biblical position and plans for maintaining a station. As mid-afternoon approached, we needed to head back since it is not safe for foreigners to travel after dark. My guide says, "I'm not going back with you. I've met some friends here that I haven't seen in a long time." I say, "But…but…but…this driver doesn't speak any English." He says, "You'll be okay," and signals to the driver to head back. I have no choice, I have to go.

Perhaps you have heard the expression of "a horse heading back to the barn". The trip back is much faster. The driver starts speeding up and down these hills – actually mountains. On the way here, I'd already noticed a spot where a car had gone over a cliff into the river below. I signal to the driver to slow down. Slow down! He slows down for a little while and then speeds back up. It was like a roller coaster ride without the safety measures until about two hours on our way we go over a bad bump. "Clang, clang, clang" something has happened to the back wheel.

He pulls over and it's starting to get dark. He jacks up the left side, pulls the wheel off. There is something wrong with the brake shoe. He crawls in under the car and I see him flexing the brake hose up and down and suddenly it snaps off. As quick as lightning, he bends a portion of the line over itself and takes some wire and ties it down like a clamp. He puts the wheel back on and says in his limited English, "Let's go". I see the hills ahead and pray, "Lord keep us safe." He did. We get back to the guest house a bit worse for wear. Dusty, tired and thirsty since we had no water, I was thankful for an excellent supper in the guest house that night.

The next day I flew out and what a spectacular sight. As we were gaining altitude, the pilot came on the PA system and said, "Anyone on the left side, look out and you can see the highest mountain in the world rising above the clouds. What a sight: Mount Everest was piercing through the clouds and gleaming in the sun. My only regret was that I didn't have my camera ready.

Today, Nepal is open to Christian radio and God is building His Church in that country. Again I was reminded how God has watched over me in countless situations. I travelled those dangerous roads safely, had no infection from the blood that was plastered over me and never had any ill effects from the food I was served along the way. Praise the Lord!

11

GOD'S AMAZING WORK IN LATIN AMERICA

"Many, Lord my God, are the wonders you have done, the things you planned for us. None can compare with you; were I to speak and tell of your deeds, they would be too many to declare." Psalm 40:5 (NIV)

BRAZIL

I REMEMBER IN 1999 WE HAD TO GET A RADIO STATION ON the air by the end of June in Brazil or the license would expire. A couple had put us in touch with a ministry in a good sized town. They needed a 250 Watt transmitter that costs around $4,000. Money was being raised from several sources. In early April, the time was getting short as the transmitter had to be ordered, built, then shipped into the country and be on the air before the end of June. Our policy is "pay as you go" without going into debt. Well, by early April we still needed $265.00. "Will you at least order it now?" they asked. I replied, "No, I can't order it but our staff will pray for the funds during staff devotions this morning." Near lunch time that day Andrea, our bookkeeper, came into my office waving an envelope: "Guess what's in this?" I said, "$265.00". "No, $266, and it says use where most needed". Again we met the deadline. How God answers prayer!

VENEZUELA

LA MORITA

It was in 1998 when we took a great short term missions team from Redeemer Bible Church, Niagara Falls, to La Morita, Venezuela to work with Pastor Zabdiel Arenas. Zabdiel had obtained a license for a radio station in this town where there were just a few known believers. It was a special project in that much of the programming would come in by satellite. This was our very first station of this type. The satellite parts were to be shipped ahead of time to Caracas to the New Tribes Mission (NTM) Base across the street from the airport.

The team of eight left Toronto loaded down with equipment, radios, tools, etc. When we arrived in Caracas, we cleared customs without any difficulty. We went across the road to load up the satellite equipment and were told they had never received it. We were puzzled, but turned the matter over to the Lord: He knew where it was. NTM let all eight of us stay in their lounge since it was already about 10 p.m. and we had to be at the airport very early in the morning for the next flight.

Rising early, we caught our connecting flight and two hours later arrived in a small town in the western part of the country. We spoke to Pastor Zabdiel when we arrived to check on the missing parts. He had heard nothing. Two hours later, after a quick meal, we boarded a bus and drove out about an hour to La Morita. Pastor Zabdiel had a mission house there where we hung up our hammocks and got unpacked. We had a great supper and then retired for the evening.

Early the next morning, we head over to the radio station site. Bruce Foreman, our radio technician, notices that the two 50-foot lengths of steel tower have been dropped off on the far side of a deep valley. The plan was to put the station on one side of the valley where the town is, and erect one tower there. Then we could stretch the cable across the valley to a high hill on the other side where we would erect the second tower. There we would set up the transmitter and the antenna.

In the meantime, Pastor Zabdiel leaves to check on the satellite equipment while we work on the towers. One has to be moved down the hill, across a swamp and up the other side to where the station is. We try to lift them and realize we need more man-power. A good crowd of people have gathered to watch these strange Canadians in their "backyard". The Lord gives me an idea. Let's get as many men on both sides of the tower, like a caterpillar, pick up the one tower and roll it down the hill towards the station. It works! The men slide down the steep, slippery hill: what a sight.

As we get to the bottom of the hill, the men again pick up the tower, but the lower part of the valley is very swampy and we are all beginning to sink in the mud with this heavy weight. What to do? Pray! We get a number of boards for the men to stand on so they won't sink. The men walk a bit, then several men keep moving the back boards to the front so they can go a little farther. We inch along in this manner until, finally, we reach the other side. A hole is dug, ready to raise the big tower into place and cement it in. However, it's very heavy to lift. They block off the street. One of the men has a very long rope which they throw over the roof of the studio building and tie it onto the back of a 4 wheel drive vehicle. The other end is tied to the tower. On the word, "go", the men down below, using rakes and poles and anything they can find, struggle to lift the tower into place as the vehicle pulls the rope over the roof. Once in place it is quickly secured. Praise the Lord!

For the other tower, more men have now arrived, and with the help of a vehicle and brute strength, we raise the second tower into place. Other men at the same time are erecting a block building nearby for the transmitter. Pastor Zabdiel has hired a man living right near the new tower location on the hill to supervise the block work on the transmitter building at a cost of $4.00/day. One of our men gives him a little extra for the great work he did. While there, we hold a children's meeting each day. The block- layer attends the first meeting and gets saved. Next day he brings his wife and family and by the end of the week they all are saved. Just this alone makes the trip worthwhile. Amazing!

In the station, we need a 19" rack to mount the equipment. They forgot to get one. One of the ladies on the team notices an old scrap metal junk yard nearby and sees a 19" metal filing cabinet, the standard size for mounting radio equipment. They cut the drawers out and make the racks fit for just a few dollars. Flexibility often comes to the rescue.

By Wednesday, after lunch, we are ready to string the power lines and antenna wire. But looking up across the valley it appears we're in for a downpour. It's not a good idea to climb a wet tower in a thunder storm! I remember noticing the time: it's six minutes after one. We stop, join hands and ask the Lord to hold back the rain for another day and a half. I open my eyes and almost immediately I see the sky clearing, God again has answered prayer.

We get all the wiring up safely on Thursday morning. Now we need the satellite equipment. Pastor Zabdiel has had no leads on where it is. What happens next is another miracle. About 9:30 a.m. a pickup truck drives in asking for Pastor Zabdiel. "I have some satellite equipment for you!" It had been delivered as planned to NTM. But the man there didn't know anything about it. He just told the delivery man to take it to the radio station down the street. The station manager knew Pastor

Zabdiel and said "I have to go across country in a few days, I'll just drop it off for him."

Praise the Lord, it arrives exactly at the time we need it. By supper time it's all assembled and the radio station is connected to the satellite. It works beautifully. After everything is installed, we sit back for a moment to relax. And it's then that the skies split open and torrents of rain sweep across the countryside. Thank you Lord!

By Friday morning the rain had stopped. Pastor Zabdiel wanted us to drop radios off at a village down the Amazon tributary in an area unreached with the Gospel. We rent two long, narrow dug-out canoes for the team and load up the radios. We use upside down plastic milk crates for seats. The first canoe takes off with a young fellow operating a little motor at the back. Putt, putt away they go.

We get in the second canoe and notice this part of the Amazon tributary has quite a current. Our fellow starts the motor but as we pull away from the shore it dies. He's pulling and pulling and we are drifting out into the middle of the river which is infested with piranhas. By this time, the canoe is really rocking back and forth from the motion of our man standing and tugging on the rip cord. I look over to the far shore, and I see two alligators slip into the water and swim towards us. Prayer time! All of a sudden the motor starts and off we go down to this village. Words cannot describe the joy of seeing these villagers each getting their own solar radio. Some of our men challenged them to a soccer game. We got badly beaten but had a lot of fun.

One day, I believe that we will meet some of these families in glory. Thank you Lord for answered prayer in so many ways on this trip.

MARIPA

In July of 2001, I was able to take a special team to Maripa, Venezuela. The team consisted of 9 ladies from Peoples Church Toronto along with Bruce Foreman and myself. The goal: to install a radio station near the border of Colombia including a 70-foot tower and antenna, and to assist with the building of the radio studio. These ladies worked hard! They dug ditches, made bricks, put up a portion of the wall for the studio building all while enduring primitive conditions and sleeping at night on hammocks draped with mosquito netting. Screams would punctuate the night silence as various insects somehow penetrated the netting. One afternoon they were able to conduct a children's meeting on the main street of town. Later on, after assembling the GALCOM radios they had brought, they travelled the adjacent river to deliver them to the Indians along the route. The people were so grateful to receive the radios and marveled that they were able to hear the Gospel clearly in their own language. Shortly afterwards, the government revoked the license for this station but God had opened a window of time for them to hear. We continue to pray that Venezuela will once again allow Christian broadcasting for its people.

PALMARITA

In discussions with Pastor Zabdiel, we learn that he has the license for a station in Palmarita so we prayerfully select a team from Community Bible Church on Hwy 7 near Lucan, Ontario and West Highland Baptist Church, Hamilton. The task is to install a 250 Watt FM station and also construct a building to house the studio.

We have about 20 people travelling with us to Caracas, Venezuela and we stay overnight in a minus 5 star hotel, but tired and weary after a long day's travel we're thankful for anywhere to rest. Oh, I must mention one minor problem: the airline lost all of our luggage. In the morning, we need to catch another flight to Santo Domingo airport in the western region of the country and expect that the airline will have our bags by then. However, none of our checked baggage has arrived from Miami. We only have our carry-ons. The airline assures us that they'll get it to us as soon as possible.

We catch our connecting flight to Santo Domingo on Venezuela's western fringe where Pastor Zabdiel has a bus waiting to take us another three hours into the back country to Palmarita. Next morning we start work in the same clothes we've travelled in and slept in. For five days we only had the clothes on our backs. The men slept on hammocks outside while the two ladies slept in the room used by the chickens. Every day we shook the sand out of our clothes and bedding that had been blown in by the unrelenting winds. We learned a great lesson as we realized that many people in the community had only one set of clothes – ever. We started work putting in the radio station, building the studio, scraping, priming and painting the tower. On the fifth day a big stake truck rolled in, horn beeping. Our suitcases had arrived. A simultaneous cheer went up from the team. Nothing was lost. Tools and equipment that we needed arrived at just the right time and everyone appreciated the blessing of a change of clothes. Praise the Lord!

Also on the property was a fenced-in area where Zabdiel raised alligators. At night, shining a flashlight into the area revealed countless pairs of red eyes. We certainly didn't swim in that river but did learn to appreciate the delicate flavor of the meat. In the remaining five days we were able to complete the project. We did have one minor incident. On our very last work day, Kurt Davis, our Board Chairman, got a nasty gash from a steel stake which required stitches. Where do you go when you're out in the jungle? We were directed to a small town nearby with a tiny medical clinic. Kurt was amazed at the way this young doctor did an excellent job of keeping everything sterile, stitching the wound neatly and bandaging it with care. It took a good while to heal but there was no need for serious follow up.

Sadly, this was another station shut down by the incoming Venezuelan government. Only the Lord knows why. Rebels came in to the station later to steal, loot and kill. Pastor Zabdiel was able to take down the tower and pack up the equipment, securing it in a safe place. He lost his property to the rebels and had to flee, but his life was spared. Pray for Venezuela and the spiritual needs of the people there.

PERU

Over the years we've worked with Felix and Patsy Liclan in Peru. Felix has now gone on to be with the Lord. One of the pastors who worked with them, decided to head downstream to a village he knew existed there. His plan was to make friends, show them the love of Christ and share the Gospel with these people. Imagine his surprise when he discovered that over 98% of the village had already come to faith in Christ. Apparently, one of their men had received a GALCOM radio and the other villagers had been keen to hear what the radio had to say. Somehow, the man was able to get hold of a speaker and managed to hook it up to the radio. Now the entire village was able to listen in. As they continued to listen, they learned how to live according to God's Word. The village was transformed and they realized they had a responsibility to reach out to the neighbouring villages with this life-changing message. The visiting pastor was able to encourage them in their faith but came away encouraged himself as he saw the power of God's transforming love evidenced in these people because of our little GALCOM radios.

On another occasion, I visited the Sacred Valley of the Incas high in the Andes Mountains.

A Galcom radio is a treasured possession for this Inca mother

It was such a privilege to be able to distribute the solar-powered, fix-tuned radios in this area and reach many of these lovely people. Although as a special guest, I was served roasted guinea pig for lunch that day, the most precious time was when I was served communion by these dedicated believers.

ECUADOR

Here is another amazing story of sacrifice and victory. GALCOM had helped the Good Shepherd Radio Station in Ecuador with radios for the Quechua Indians. They reported the following:

One morning a local shoemaker was running the radio station as a volunteer at 6:00 in the morning. A Quechua Indian with his wife and family came into the studio. The Indian man explained that they had travelled a great distance from a remote village. This village had only one GALCOM radio which hung in a tree so the entire village could gather around and listen to the Gospel broadcast in the Quechua language. His family along with other villagers had been listening to the program for some time. They took up an offering for the radio station which he and his family had come to deliver. It amounted to just under $1.00. He and his family had walked 10 hours to a bus stop, then rode the bus for four more hours. After he presented the offering to the shoemaker, he asked how he and his family could become followers of this Jesus. The shoemaker put on a long-play album for the morning broadcast, took the family outside and led them all in a prayer of repentance and commitment of their lives to Jesus Christ. The family left full of joy. They had come to bring a gift but left with the gift of eternal life!

I was also privileged to travel in among the Waodoni Tribe (often known as the Auca Indians or Huaoranis) who had in 1956 killed five missionaries who attempted to befriend them: many of you will be familiar with that story from the Books, "Through Gates of Splendour" and "End of the Spear". Many of these beautiful people have come to faith in Christ choosing to follow "God's Trail". Among them are Mincaye and Temente who travelled from Ecuador to Canada to share the story of what God is doing among their people at our Annual Banquet.

Allan flew in with Steve Saint to meet Mincaye & Temente

Mincaye was one of the men involved in spearing the five martyrs. We are still hoping to set up a radio station among them to reach into the other Waodoni settlements. On our flight out, we experienced such thick fog that Steve Saint, with only 20 minutes of fuel left, had to make an emergency landing at an unused military "air strip". It took some explaining looking down the barrel of a rifle to explain the emergency to the agitated soldiers. Unable to find fuel, a ten hour bumpy bus ride eventually found us back in Quito.

BOLIVIA

DEVIL'S ISLAND

Beth and Ricardo Chavez were serving as missionaries in Sopachuy, Bolivia. Sopachuy means "devil's island". They say the name is well chosen because it is said this town has the lowest percentage of believers in the whole area. This young couple labored for years with determination and prayer. The Holy Spirit began to work in hearts and a number of the people came to know the Lord. However, living in a remote area with poor roads and homes scattered all over, Beth's dad, Jim Driscoll, suggested they contact GALCOM. As a technical person he was aware of GALCOM's unique ministry.

As they communicated with us we laid out plans for a radio station. With Jim's help and the financial support of their home church, Calvary Chapel, North Carolina, they were able to set up an FM radio station in the town and we provided them with 500 radios. This community has now changed tremendously from darkness to light, from hopelessness to happiness and from bondage to freedom in Christ. Beth and Ricardo wondered at one point why God had placed them there but now they could see the transformation taking place before their eyes. Devil's Island had become God's haven.

QUECHUA INDIANS

The distressed man was wandering on a lonely road up a 15,000 foot mountain when Alex and Judy Muir, of Pioneers Canada, met him. The cause of his agitation was a broken radio. He seemed to be acting rather strangely and kept saying, "Fix it; it's my life!" apparently, a missionary from Georgia had given out 25 radios tuned to "Mosej Chaski" ("New Messenger") short wave radio station. Alex took the radio and a few days later just happened to meet up with another missionary who was able to repair the man's radio. Alex and his son, Greg, returned to find the man. He literally jumped two to three feet in the air, he was so excited. Alex and Greg figured that if these radios were having such an impact they needed to find out where to get them.

Since the GALCOM label had a Canadian address, Alex lost no time on his return to Canada to contact us. By October 2004, he had enough money raised to order 300 GALCOM solar-powered fix-tuned radios. This quickly doubled to 600. By April of 2005, Alex had enlisted a team of nine people to travel into Quechua territory in Bolivia for a distribution campaign.

The radios were so treasured by the Quechuas that it was impossible not to be elated by their response. Mosej Chaski consistently taught them about a way of peace, truth and eternal life through Jesus Christ. That mission trip caused Alex to change direction from construction to radio distribution, and an ambitious vision rapidly formed of taking two teams a year with a thousand radios each trip for the next five years.

To date, Alex and Judy with 25 teams in seven years have handed out 37,300 GALCOM GO-YE fix-tuned radios. As Alex remarked, "Without a doubt, over a million Quechua people are under the sound of the Gospel. For every home that possesses a radio, the homes on either side willingly hear the message. Over 20 churches have been planted in this area with GALCOM radios as their pastor."

Tim Whitehead, our executive director, went with Alex on one of these trips in 2008. First, Alex explained, they needed to act as burros to carry the radios into the country, since taxes and duty made it too expensive for shipping. This may sound fairly simple, but trekking up into the high mountains where air is thin can be quite a challenge. As Tim would say, "You can breathe, but it doesn't do you much good." When they reached the designated area, they would have to attract the people. Many would hear the gringos were coming and walk for days to get a radio. The team would sing songs and play with the children. Soon, a large crowd would gather. They would carefully distribute the radios, one per family and then move on to the next area. Now that we have the ImpaX radio which includes the Quecha Bible, the people rejoice at now having God's word as well in their own language.

Another task the team assumed was to encourage the few pastors and missionaries that they met along the way. Most of the Quechua have very little education (although many are very intelligent), financial support or contact with the outside world.

Thousands of testimonies could be included here but space limits us to one. One evening, in a Quechua village, a boy came running to Greg saying, "Come, come". They went into a small home where a witchdoctor was listening to a GALCOM radio. Excitedly, the witchdoctor declared, "There is no other god except Jesus Christ." Right there, laying aside all of his enchantments, he committed his life to the Saviour. We will never know how many others were influenced by him to turn to Christ. This we do know: drunkenness, drugs, fighting and immorality have plunged as the people are finding their true purpose and life in Jesus.

PARAGUAY

ASCENCION

Also in Paraguay, I had met Pastor Holowaty in Ascencion and quickly recognized he had a great heart for reaching people with the Gospel. He very specifically asked for AM radios with solar panels and built-in speakers to distribute in the jungle villages. We sent him several hundred radios initially which he carefully distributed in this virgin territory where, to his knowledge, there were no known believers. Two years later he took a church planting team into this area and to his amazement over 250 people showed up, all new believers because they had been listening to the Gospel over their GALCOM radio receivers. They were eagerly waiting for a church to be started. In Pastor Holowaty's experience he had never seen a church planted so quickly.

Prisoners are another group Holowaty has a special burden for: those who are not only imprisoned physically but are also held captive by their sin. Holowaty worked diligently to arrange for GALCOM fix-tuned radios to be given to each prisoner. The women's prison was opened first. After only two weeks, the warden called: "I have 33 ladies wanting to be baptized." Well, that was the first for the prison; Pastor Holowaty was allowed to baptize all 33 within the prison walls. One day we will meet them in heaven.

He went to another prison called Panchito for boys under 18 years of age. These kids were fascinated by the little radios and in no time at all, 18 boys requested baptism. Holowaty then went to San Juan Bautista, a maximum security prison 120 miles outside the capital city of Ascencion. Very soon over a dozen men turned their lives over to Christ. We praise God for the involvement Pastor Holowaty has had in prison outreach.

NARANJAL

It was a hot, humid, five hour drive from the airport in Ascencion, Paraguay before Dave Casement and I arrived in the jungle area of Naranjal to install the first Christian radio station for Pastor Ramon Aguero only to learn that our transmitter equipment had not cleared customs and was still in the city. This, after extensive negotiations with the courier company before leaving Canada promising that they would have it there for us. This presented a huge challenge. What do we do? The only option was to leave Dave to start on the tower while Pastor Ramon and I drove back to Ascension. We arrived at the courier office just before closing time. "It will take a week," they tell us. We walk out discouraged saying, "Lord, we need your help."

By "chance", just as we are leaving the building, we meet Pastor Holoway who runs the Christian radio station in the capital and tell him our situation. One of his workers says, "There's a small one-man custom broker nearby. I think he can get the equipment for you." We stayed over at Pastor Holowaty's guest house and prayed late into the night. I remember waking up early the next morning about 5:00 a.m. and looking across the room. There is Pastor Ramon on his knees beside his bed crying out to God for help. I say, "Thank you, Lord, for such a man of faith and commitment."

We are at the little customs office as soon as it opens and led into a one-room operation with a desk piled with papers. The man looks to be in his seventies: behind another desk he has a secretary doing paper work. It only takes 15 minutes for the man to check over the papers. He raises his head and tells Pastor Ramon, "Come back at 2:00 p.m. Your stuff will be ready to pick up." We look at each other in amazement. Here the huge courier office with all kinds of staff couldn't do anything for a week and this one small operation can do it in a few hours. We had real peace as we left the little office. Back at two, he hands us the cleared documents to pick up the equipment. Praise God from whom all blessings flow! We arrive back in Naranjal about 9:00 p.m. tired but rejoicing in the Lord. Within a couple of days the station was on the air and the radios distributed to the people. That station is still blessing thousands of people in that corner of Paraguay.

From the time I first met Pastor Raymon at the airport, he seemed to have a strange walk as though he were dragging his right foot. As we sat in the customs office waiting to free the transmitter, Pastor Raymon crossed his legs and immediately I realized why he was dragging one foot. The sole of his right shoe was almost completely off. In order to keep it from flopping around, he had to drag his foot. It just so happened as we headed out for lunch that we passed a shoe store and soon our brother was taken care of. You would think we had given him a million dollars. I learned later that he and his wife give so much to the villagers that they scarcely have enough for themselves. Again I praised the Lord for servants like these. As he has travelled into remote villages distributing GALCOM radios he has seen a great harvest of souls. How often I have remarked that there will be many "nobodies" in the eyes of the world who will be in the front row in heaven.

COLOMBIA

"And the work of righteousness shall be peace; and the effect of righteousness quietness and assurance for ever." Isaiah 32:17

COLOMBIA TEAM MINISTRY

Working with other ministries has been one of the great blessings over the years. We have been able to take many teams to the mission field to work on radio station projects and other related projects that will help people hear the Gospel. One of these trips was in January 1995 when Dr. Julio and Ruth Ruibal invited Florrie and me to bring a team to help with their ministry in Cali, Colombia. The team of 21 people from Philpott Memorial Church, left Hamilton on January 21, 1995 and arrived at Cali where Julio met us and whisked us through customs in about ten minutes. No cases were opened and no charges were made on the equipment we were bringing in. This, in itself, was a testimony to the tremendous respect many locals had for Dr. Ruibal.

Cali is one of the most violent cities in Colombia and during the three weeks prior to our arrival, there had been thirty nine murders. The Ruibals had an active Christian ministry including a church, school, medical centre, training centre for pastors and radio ministry. We came for eight days to put in three new Christian radio stations as well as do a number of tasks around the complex such as painting, carpentry, building a radio studio and outreach ministry.

We divided into three groups: one to do painting, carpentry and minor repairs in the school, another to work on building the cement block radio studio, and a third group to install two radio stations in other locations. Praise God we were protected and much was accomplished.

One of the most exciting parts of this trip was to install an AM radio station for the Guambiano Indians near Sylvia, Colombia. These people had been neglected for years, basically living off what they could grow in the mountains, supplemented with a few chickens. Very few had heard the Gospel. One evening Dan Shaw, Derek Wilson and I accompanied two of the Guambianos by truck up through the winding mountain roads for about a three-hour drive. As we arrived, Pastor David, one of the Guambiano leaders who had just been trained by the Ruibals in broadcasting, came out to welcome us. Pleased, he beckoned us, "Come see our broadcasting studio." They had cleared the chickens out of the former hen house and set up an old white table to hold the transmitter and studio equipment. The hour was late so we went to bed hoping to get an early start next morning.

The Guambianos are a very small people, even the men are barely over five feet tall but man do they know how to work! In a few hours we had the four tall poles raised to form a 90,000+ cubic foot area for the antenna and were able to attach the positive wire to the tops of the poles; this was no small feat on a steep mountain side. By noon the antenna was mounted and installed. Then Dan supervised digging the trenches for the ground plane using gestures and a Spanish dictionary in hand.

Derek Wilson & Dan Shaw assist the Guambianos in the radio station installation

How do you explain that you need eight radial trenches dug from the centre of the square enclosed by the poles stretching out about 325 feet? Tricky, but they soon caught on and the copper wire was laid in the trenches for the ground plane.

Meanwhile, the ladies were preparing a nutritious lunch. I set up the studio in the converted chicken coup and by early afternoon everything was connected and ready for broadcasting. Now for the test! Oh, the excitement as Pastor David in the Guambiano language, began to speak over the many little GALCOM radios that we had distributed to the people. The signal was loud and clear and everything worked perfectly. The tears came to Pastor David's eyes as he realized the impact this radio station would have on his people scattered throughout that mountain range. Outside the little studio, joining hands, we thanked God for making this a reality and seeking his blessing on the broadcasts in the future.

Since January 1995, that station had touched thousands of lives. I believe over 30,000 Indians live in the area covered by that first station. Since then the station has been upgraded to 500 Watts reaching over 100,000 people. Only eternity will tell the final story. To our surprise, in 2008, Pastor David came to our Hamilton office and brought in a carved wooden plaque attesting to the effectiveness of the radio station on thousands of Guambianos.

The national pastors, Dr. Julio and Ruth worked diligently to promote huge prayer rallies seeking to bring unity to the churches in that area. Sadly, only a few months after our visit there, Dr. Julio was assassinated as he went to meet with local pastors for that purpose. His wife, Ruth, soon took over leadership of the ministry which has continued to expand since that time.

GOSPEL OF PEACE

God had given Russell Stendal, of Colombia Para Cristo, favour with a number of high ranking officers in the Colombian military so that he was regularly allowed to distributed radios, Bibles and Christian literature to the soldiers, the effect of which was apparent in their demeanor. Russell was also greatly esteemed by many of the guerillas who had previously held him in captivity for almost half a year under armed guard where he faithfully shared the message of Jesus Christ and lived to write a book called, "Rescue the Captors".

In 2001 I was privileged to meet up with Russell at Mission Fest in Vancouver. We shared our ministries with each other and both of us were very excited about the possibility of partnering together. Russell already had several Christian radio stations broadcasting in Colombia. We had the means for people to hear those broadcasts – the GALCOM solar-powered fix-tuned radios. Shortly after that meeting, we were able to send out the first 1,000 radios to Colombia. Russell began distributing them along with Bibles and books.

Then in 2004, Russell came to us with an ambitious proposal: The civil war in Colombia had been raging for decades. Corruption and drug trafficking were rampant. At any one time, between the Colombian army, paramilitary and the Marxist guerillas, there were upwards of 1,000,000 people actively involved in the fighting. Russell proposed that if we could supply him with 100,000 radios over the next five years, that would allow for approximately one radio for every 10 people involved in the conflict. Russell was convinced that if all these people heard the Gospel and the true way of peace, that God would change his country.

We were a little overwhelmed. In discussions with our major funder, Harold Kent, through GALCOM USA, and Gary Nelson who administered those funds, several things were clear. First, the need was a pressing one. Secondly, Colombia Para Cristo was a well-established, stable, courageous, evangelical ministry. Thirdly, Russell's track record showed he had the capability of distributing such a large quantity of radios. Fourthly, with two powerful short wave stations, he was able to cover a large portion of the country. Fifthly, this would require huge funding at $20.00 U.S. per radio. Sixthly, we needed to do this without jeopardizing the supply of radios to the rest of the countries where we were involved. And finally, we needed to be sure this was the Lord's leading before embarking on such a big venture.

At our 2005 U.S. Board of Directors meeting in Tiberias, Israel, by faith, it was agreed that GALCOM USA would fund the radios, GALCOM International, Canada would manufacture and ship the radios and Russell would strategically place them for the greatest outreach.

Because of the network of contacts he had established at that time, copies of his book, along with the radios, were welcomed into the various guerilla camps. Then, to reach many of the eastern areas of the country Russell started requesting small parachutes. Without proper roads much of the country could only be reached on foot, by horse or donkey or by river all of which are extremely slow and cumbersome when carting thousands of books, Bibles and radios. By flying his Cessna over the targeted areas, Russell was able to make many parachute drops in one flight.

Russell Stendal prepares his airplane for another parachute drop

Thousands of radios have now been distributed in Colombia using this method.

Month by month, GALCOM USA provided funding for 5,000 radios. Radios poured off the production line in Canada. Russell, along with his team, working in life-threatening situations, faithfully continued the task of distribution. Above all of this, God was working. Slowly but surely, attitudes in Colombia were changing. Many people were beginning to acknowledge that war was not the answer: God just might have a better solution.

As the radio broadcasts went out hour by hour and day by day, thousands of Colombians were turning to faith in Christ.

Volunteers pack parachutes they have made for Colombia

The Gospel crossed all barriers: Colombian military, paramilitary, guerillas, farmers, common folk. In some localities, hostilities began to weaken. Russell's daughters, Lisa and Alethia prepared a full length, award winning documentary film called "La Montaña" describing how God melted the barriers on this high mountain and turned it into a transmission site for the message of peace in Jesus.

COLOMBIA MYSTERY TRIP

As thousands of radios were being distributed into the country, Russell invited me to see the ministry in progress. I arrived in Bogata with a team mate, Joe Liete, who had arranged our flights. It's about 11:30 p.m. when we arrive at Russell's home. He greets us and says, "I've just been called to visit a friend of mine who needs help but it's in very dangerous guerilla territory. You can wait here until I return or come with me. But we have to leave by 3:00 a.m." I knew God was leading Russell to go; he would be under God's protection and that was good enough for me. Russell's wife, Marina, and her brother join us as well.

After about six hours drive, we arrive in a remote bush area and Russell stops by an old barn. Out come two men from behind the barn with three donkeys and they load our personal items on the donkeys along with the GALCOM radios and off we go. Russell's wife tells me we now have about a two and a half hour walk into the mountains. We start trekking and arrive at a place high up on a hill overlooking the area. I survey the beautiful mountain peaks, lush valleys and rippling streams. The peaceful surroundings belie the tense, life-threatening battle waging fiercely under its green canopy.

As we approach the camp, the guerillas welcome us. To our knowledge we are the first Canadians to be allowed among these people. The camp consists of about two dozen men, a mother with a child and an older lady all interested in who we are and what we do. So Russell tells them about us. Then, one of the men, through an interpreter says to me, "You work on transmitters? Ours is broken. Can you fix it?" Well, you don't argue with someone holding an AK47, so I go behind the building. There it is; they start the generator but there is no life in the transmitter. I have no tools or meters with me, but I pray and then ask for a Philips screwdriver. Removing the top off the case, I see the fuse block. "Lord, how can I test this without a meter?' An idea comes to my mind and I ask for a flashlight. I pull out the fuse, place it on the battery which I have removed from the flashlight, put the light bulb on top of the fuse block and with a nail I complete the electrical circuit. The light flashes on. I repeat the procedure with the other fuse: again the light flashes on. Then I uncover the next area and discover a burnt coil. Without a replacement on hand, Russell will have to take it back to Bogata and get it repaired. The group of men running this transmitter are allowing Russell to provide Christian programming to be aired in that region – that in itself is amazing. That is why I sincerely want to get the transmitter fixed.

By now, it's getting too late in the day to make the long trek back to the vehicle. Russell had wanted to get in and out of the area as quickly as possible, but now he informs us we'll have to stay overnight. We're all a bit uneasy about this but we have no choice. Supper is a small hunk of beef (I think!) in a bowl with some vegetables, broth and a spoon. The guerillas watch us discreetly to detect our reaction to their primitive life style but Joe and I dig in with the rest of them and enjoy the meal. They're pleased to see that the "gringos" are not too proud to eat their simple food. It gets dark early there and we're outside at a picnic type table on the porch, some on chairs and some sitting on the rail.

One of the men asks Russell to ask me to tell them something about myself and Canada. With Russell interpreting, I share how God has directed in my life, then I share a bit about Canada and why we do radio work. The setting is eerie, only one candle is lit on the table in the pitch black. All I can see are the whites of their eyes fixed intently on me, taking it all in. Tactfully, I present the Gospel to them and leave it to the Lord to work in their hearts.

It's about 10:30 p.m. now and time for bed. Joe has a bed to sleep on but I am given a thin rubber mat on the floor with a car blanket, one of those with little tassels that tend to tickle, to cover me. My jacket makes an adequate pillow and I fall asleep.

Although the mountain air is cool, I wake up about 2:00 a.m. in a sweat and I sense spiritual warfare. A voice is taunting me, "Allan, you are going to die. You will never see your family again. You are a fool for coming here. They are going to torture and kill you or hold you for ransom." This they could easily do and no one could stop them. In a low voice I pray out loud that the shed blood of Jesus will cover and protect me from the power of Satan. Moments later, a peace sweeps over my heart and I fall back to sleep. Thank you Lord.

I get up about 6:00 a.m. and head outside. The sun is just beginning to shine – what a beautiful sight. To wash up, there's an old pipe coming out of the mountainside feeding water into a rickety tub which is overflowing and creating quite a mud hole. Man, that water is cold; it certainly wakes me up. In a little while the others are up and by mid-morning we are on our way home. First, Russell needs to have a private talk with one of the men he's come to meet. Eventually we start out expecting Russell to catch up to us. He doesn't show up and we start thinking of all the possibilities of what could be happening. After a long wait, much trepidation and many prayers, Russell finally appears up the trail and we continue on our hike to the vehicle.

It's getting on in the afternoon when we start to drive through a long tunnel that cuts through the side of the mountain. It's the only road to Bogata but a big truck has broken down in the tunnel and traffic is backed up including us. Russell's getting a bit worried because if you get stuck out on a remote road like this at night, anyone could hold you up and take whatever they wish. No one in the other cars would help, they would just back away leaving us on our own. Many drivers are held up and/or shot so we pray for the Lord's intervention. Time passes and we're beginning to feel the pressure but just as the sun is receding, some policemen on motorcycles come along and pick out certain cars including ours and escort us through the tunnel, around the truck blocking things, and soon we're on our way. We return to Canada again praising the Lord for answered prayer and for the opportunity to witness to the people in the mountains. What a great God we serve!

COLOMBIA'S 100,000TH RADIO

July 2012, the time has finally arrived – a little behind our planned schedule, but the 100,000th radio flows off the production line. In conversations with Russell Stendal it's agreed that this momentous occasion requires a celebration. Gary Nelson, president of GALCOM USA and Kris Mineau, a GALCOM USA Board Member are both retired air force colonels. The three of us will travel to Bogata, Colombia and then on to Cali to present this 100,000th solar-powered fix-tuned radio at a special ceremony to Colombian army lieutenant General Barrero at his headquarters. Gary will do the honours, while I present the very first GALCOM ImpaX player (combined radio and audio Bible) to his deputy, Major General Segura. Kris Mineau offers a prayer for God's blessing on these two high ranking Colombian officials.

Gary Nelson presents the 100,000th Galcom radio to General Barrero

During this visit to Colombia, we have the opportunity to meet José (not his real name) a former guerilla leader who now has his own Christian radio station. Thousands of guerillas are turning to the Lord, thousands of men and women throughout the country are embracing the One who is the Way, the Truth and the Life. Estimates reach as high as 30% of the Colombian army becoming Christians. God continues to work in this troubled land. The following three testimonies give some inkling of what is happening.

PABEL

I am forever indebted to the Lord. I lost my sister when she was only 16. We did not know that her boyfriend was involved in terrible things until one day we found her in his house poisoned with an insecticide. Strange people approached me saying that I could get revenge by throwing a grenade on his house and doing away with them all. But I told them that I would leave it in God's hands and that He would work justice. A year passed, and I heard that he and his brother had been shot to death by the guerrillas because it was rumored that they were part of a violent gang that stole motorcycles.

The boyfriend's family thought that the deaths of their sons had been caused by vengeance from my family who had sent the guerrillas to kill their sons. I was serving in the military then. We were in serious trouble. Even though we didn't know the Lord at that time, no one in my family had sent the guerrillas because we had left everything in God's hands. The boyfriend's family hired 30 armed mercenaries to kill my family. They started by shooting my cousin in the head. After that, everyone in the village began to flee, and others (including my parents) couldn't sleep because they were afraid. Terror had overtaken the village. People began to tell me that I had to flee far away. I knelt down in front of my house weeping and prayed to the Lord in front of my family saying; "Lord, I am going to trust you. I am not going to run away and I ask You to send Your angels to get these people that are causing so much damage out of here". The Lord heard my prayer and saw my tears. Fifteen days later 200 paramilitary came and forced the hired killers to flee.

The most special day in my life came that I will never forget. I was in my house making an antenna of aluminum tubes trying to hear a shortwave radio station better because I wanted to learn English. As I was climbing the tree to connect the cable I heard an explosion in my house and I thought it was my old green radio. Instead of a burnt out radio, I heard for the first time in a very clear way Russell Stendal speaking. My friend Jimeno had doubts about religion. I ran to him with the radio and said, "listen! listen! listen! This is different! I haven't heard this before!" Jimeno and I listened in complete silence when he looked at me and said, "This is the truth!" And the radio program was called "The Truth about the Truth." We were like kids beginning to walk in this new way and we stumbled and fell and we limped until we were able to walk straight as is God's way. All of us who have seen the GALCOM audio Bible along with the radio are delighted. And we don't know what to do since anyone who sees it immediately wants me to give them one. Thank you so much for the Galcom Bible radio. Jimeno almost cried of joy when he received his and he can't wait to listen to it.

ONE RADIO, ONE MILITARY BASE

One day a mother asked for a radio for her son, a professional soldier, who was fighting guerillas on a high mountain military base. The sergeant in charge saw the radio and asked, "Where did you get that?"

The soldier responded; "From my mother."

The Sergeant said; "I know those radios because I used to have one. Its messages are wonderful and I gave mine to my mother. I'll pay you 100,000 pesos (the equivalent of 50 dollars) for that radio."

The soldier replied; "I can't because my mother told me to listen to it. She said it would be my protection."

The Sergeant answered back; "Soldier, I'll give you an order: at 12 p.m. and 6 p.m., you will turn on the radio and all the soldiers who are off duty will sit and listen to the messages. It's an order!" That's how God's Word reached the men in the mountains of Northeastern Colombia. The soldier's mother later told me that more than 70 % of the men in that military base now walk with the Lord.

ONE REMOTE VILLAGE

Tomas, a professional house painter, came to the small village of La Paz (The Peace) looking for a job. He was extremely curious to see a little green radio on the railing of a poor house. Since he had been distributing these GALCOM radios in other areas of the country, he wondered how this radio, had made its way to such a remote location where there has been strong fighting between guerillas and paramilitary for the past several decades. So he asked the owner, "How did this radio get to you?"

The farmer responded, "A missionary gave them away in another town and we got five radios." This little village of La Paz had about 80 people. The villager continued; "As you can see, no one ever comes here. The priest doesn't come because it's too far out so we're alone. We rotate these radios along with three Bibles among the villagers. We've been listening to these messages for months. A brother who was born here gives us a message on Sundays through the radio. We don't have a pastor for ourselves. No one has come to evangelize us. Only the Lord has evangelized us through the GALCOM radios. Everyone in this village follows the Word of the Lord!"

12

A GLIMPSE OF AFRICA

"So is my word that goes out from my mouth: It will not return to me empty, but will accomplish what I desire and achieve the purpose for which I sent it." Isaiah 55:11

GOD'S BLESSINGS IN SIERRA LEONE

Another of the great blessings that I remember was when we sent Dave Casement over to Bo, Sierra Leone in 2003 to install an FM radio station and donate 2,000 Galcom solar fix-tuned radios. What an opportunity in a Muslim country. Seven years after the installation of that radio station approximately 37,000 Muslims had come to trust Christ as Saviour along with 15,000 non Muslims. Over 970 new churches had been planted. This calls for a special Hallelujah!

BURKINA FASO

It was December 31, 1999, New Years Eve. Apparently not too many people wanted to fly that day, especially overnight because of the Y2K scare. Many of you will remember the horrific predictions of all kinds of malfunctions in computers due to the changeover to the twenty first century. I got a super price from Air France on a ticket to Burkina Faso, West Africa, to install two radio stations in Ziniare and Leo. 2,000 fix-tuned radios had previously been shipped to the radio station in Ouagadougou, the capital: 1500 for Ziniare and 500 for Leo ready for distribution as soon as the stations were completed. December 30th I flew to Paris, stayed overnight and then flew out the next morning to Ouagadougou.

Burkina Faso is a country of over seven million people mostly Muslim in sub Saharan West Africa. French is the official language and there are five major languages with dozens of dialects. Praise God, He was allowing us to place two more complete stations in remote communities. Arriving at about 4:00 p.m. I stepped out of the plane into the searing desert heat. Pastor Michel met me at the airport, bought me a Coke, and drove me to the mission guest house where I would be staying. He said to get some rest and he would pick me up around 9:00 for the church's Watch Night Service celebrating the incoming New Year. I go to my room: there's a big double bed with one sheet, no blanket and a huge sofa cushion for a pillow and very few amenities in the bathroom. Everyone seemed to be away for the holiday or busy with activities. No one was in the kitchen preparing food so there was no supper that night but surely refreshments would be served at the New Year's Eve service. About 8:45 Pastor Michel's horn beeps and I join him for the ride to the church. There must have been over 500 people, many converted Muslims but no food. The service goes on till about 1:00 a.m. and I'm trying to stay awake: my body says it's sleep time and my stomach is growling. Back in my room I'm thankful that my wonderful wife has packed an abundance of granola bars, assorted nuts, raisins and other dried fruits so I snack before hitting the sack. Pastor Michel told me breakfast would be served at 8:00 a. m. next morning. The evening air is cool so I use my coat as a blanket and settle down for the night.

The next morning, I head down to the kitchen along with a couple of American missionaries who have arrived. The middle aged chef greets us and says, "Sorry, no breakfast today; there's no food; come back at 12:00 for lunch and if no one is here, help yourself. By noon I'm pretty hungry so I head down to the kitchen where I meet one of the Americans. No one is there so we open the refrigerator to see an old dried up cabbage and a bone off some animal. We look around the kitchen and find nothing more. Checking out the back door we see a dried up goat's head on the ground and a boy across the way filling up plastic water bottles with water to resell – not something fit for a North American stomach.

About 4:00 p.m. the Pastor arrives. "How are things going?" He asks. I sort of hint that the food is pretty scarce. He remembers it's a holiday and apologizes that I was not looked after. Right away he takes me to a roadside café. I look over the French menu with my limited French. My policy in these countries is, if you can't cook it or peel it, don't eat it. I read what I think says "chicken noodle soup" and order that. Pastor Michel orders barley soup. When the soup comes, mine contains very fine noodles like thin string and on the top are floating several small round orange rings. I ask Pastor Michel what it is; he calls the waiter and then announces that's chicken feet soup. Suddenly, I lose my appetite. Right away the pastor says, "I like that," and switches bowls. I'm much happier. Then the Pastor says, "That's not much to eat: you need more than that." He points to the trees and says, "Look up in those tall trees: see those dark images hanging from the branches? They're fruit bats and their meat is delicious." He calls to the waiter, "Give us two orders of fruit bat". I say to myself, "I can't believe this." The waiter responds, "I'm sorry, we have no bats and because it's a holiday there's no one to climb the trees." I'm relieved.

Later, after visiting Pastor Michel's family, I'm back in my room and preparing myself to start work on the radio station the next morning. Only two days have been allotted for this installation. I know that the transmitter and all the studio equipment have been shipped in ahead of time so I'll only have to be concerned about transporting a few tools along with my personal effects.

This area of Ziniare is arid desert with continuous blasting sand. The people are desperately poor and mostly Muslim. A 2,000 Watt station here should reach tens of thousands of people across this flat expanse. They'd already done a terrific job of erecting the 150 foot tower and had built a nice mud brick building for the studio complete with enough furniture to hold the studio equipment. Amazingly, they had reasonably reliable electricity. To my chagrin, as I entered the studio, I was met with a very shocking sight. They'd unpacked every box and every part: nuts, bolts, clamps, wires, antenna parts, studio equipment were all spread out all over the place in a disheveled heap. Talk about a jigsaw puzzle! I thank the Lord for answered prayer in enabling me to get everything sorted out again.

Two young men were quite eager to climb the tower, for which I was thankful, so I rehearsed with them the assembly of the antenna on the ground. They quickly caught on and in no time were up at the top of the tower securing the antenna. It was somewhat amusing to have a crowd of people gather around to watch my every move and a circle of ladies following me everywhere I went.

With everything completed it was time to turn on the power. What a thrill to see all the equipment light up. I checked that all the settings were accurate. The Pastor excitedly began talking into the microphone and the fix-tuned radios picked up the signal perfectly. What a busy two days!

As we distributed more radios to the families in the area, it was exhilarating to see the expressions on the people's faces as they turned on the radios and heard the Gospel message in their own language. I went over to one old, thin man, obviously very poor with just a blanket wrapped around him and handed him a radio. I thought he was going to cry he was so happy to receive it. It's quite possible we will meet this man in heaven.

A blanket to warm his body and a radio to warm his heart

After a prayer dedicating the station to the Lord, I moved on to the city of Leo. Another interesting note here is that we averaged about one meal per day supplemented with bread and cheese for breakfast. Again, I was thankful for the extra snacks Florrie had packed for me.

Leo at least had a few trees growing and slightly better living conditions.

I had expected to start from scratch with this station but a young married man from the area had helped some HCJB radio people in the past and to my surprise, he had the new station completely set up. I checked over all the settings and everything was accurate. I was so proud of him as we prayed for God to bless this station. Giving out the GALCOM radios is always so special. To see the excitement and anticipation on the people's faces is always a beautiful part of the task. God has said that His Word will not return to Him void but will accomplish what He has sent it to do. We are trusting the Lord for another great harvest of souls from Leo and the surrounding area.

2,000 Watt transmitter ready to assemble and install in Burkina Fasso

I headed back to Ouagadougou and was invited by Fred Hastings with the Christian Blind Mission, for supper with his family. He and his wife were working in the area because many of the children are struck with blindness at an early age caused by onchocerciasis, or river blindness. Radios for these children mean they can still stay connected educationally as well as hearing the Gospel. They love their little GO-YE radios. Praise God for what He is doing in this country.

I was made a bit nervous when I was about to leave and the leader called me over. As I stood beside him, he pulled out a sword from its sheath. It must have been about three feet long. I wondered, "Oh, oh, is this the end?" No, rather, it was a beautiful gift for all the work I had done. It had a long curved blade with many ornate carvings on the sword handle and the sheath. Arriving at the airport, I am wondering what will happen to this sword I am carrying under my arm. I wasn't in suspense for very long. As soon as I entered, a soldier pointed to my sword and signaled that he wants it. So I said to myself, "That's the end of that." But no, he follows me to the Air France ticket gate and hands the sword over to the clerk. He puts a tag on it with a bit of plastic wrapping around it and throws it on the conveyor. Again, I am convinced I will never see it again. However, at the Toronto airport up comes the sword on the baggage conveyor. The sword, now on my office wall, reminds me of the two new radio stations on the air and the several thousand fix-tuned radios which saturated that area of Burkina Faso. In Pastor Michel's church alone there must be over 1,000 people who now have come to know Jesus through the solar-powered radios. Many of them are former Muslims. Praise God for what He is doing in that country.

UGANDA

We had a couple in to the office just back from Uganda with testimonies and pictures showing how the radios have been such a blessing to the people there. They provided us with this testimony from a Ugandan contact:

"Greetings in Jesus' Name! I am happy to testify what the Lord Jesus has done to me. I am happy for the solar radio which was given to me. This radio is tuned to a Christian frequency, Impact FM. This frequency has much of Christian programs, prayers, music and counseling which has given me opportunity to change in my spiritual life. Since it doesn't require dry cells, it's cheap so I use it all the time. It helps get me up since there is a prayer program at 5:00 a.m. It has helped me to wake up early and also to prepare for my work. This being a unique radio it has made me look unique where I use it. I even listen to the lunch hour program since it is portable to the place of work. Not only my life, but even other church members who were given one have been sincerely blessed. Boredom has been reduced. We thank our dear missionaries who supplied them to our churches."

13

CHRISTIAN RADIO IN CANADA

"Blessed is the nation whose God is the Lord; and the people whom he hath chosen for his own inheritance." Psalm 33:12

CANADA HAS BEEN ONE OF THE MOST DIFFICULT AREAS TO secure radio licensing for Christian broadcasting. It's only in the past few years that a few licenses have been granted. In the meantime, the policy has been much more open for first nations people and we have been able to set up several radio stations in these areas including the following two.

PICKLE LAKE – THUNDER BAY: REACHING THE FIRST NATIONS

Our goal was to have a radio station for the Ojibway and Cree nations in Northern Ontario. They have been resistant to the Gospel because of a long history of abuse in previous years. Over this vast region we're told they number around 190,000 spreading out across the land. It took three years of negotiations to get a license for these stations. I still remember Dave

Casement, Ron Bereza, Walter Dickson and Harley Smith travelling up in the middle of winter to put up the tower. They battled freezing cold wind and ice but in spite of all the challenges they completed the task. Promising inroads have been made into these communities since this station has been on the air and lives have been totally transformed. Thank you Lord.

FORT VERMILLION, ALBERTA

It was nearing December 2003 and we'd been working with Michael Sandstrom of the Northern Canada Evangelical Mission for at least two years trying to get a license for a radio station in Fort Vermillion in northern Alberta, to minister to the First Nations people there. The CRTC in Ottawa had finally given him the license just before Christmas. However, they only gave us 30 days to get the station on the air. Most governments will give a year to a year and a half. We had no choice but to quickly assemble the necessary equipment and tools to take with us. What do we do when winter sets in early especially in this challenging northern region? We just had to prepare to travel up there in the dead of winter and install it.

Dave Casement, one of our radio technicians and I fly up in the middle of January. Michael meets us and we stay at his home. The temperature is minus 45 degrees Fahrenheit.

The next day we take off to the radio station site. Assembling the studio equipment inside the new studio building is no problem. But for the tower they're renting space on a Telus cell phone tower and only the Telus men are allowed to climb it. So we have to have everything ready: the antenna, coax cable, clamps, tie downs and the cable measured out exactly for them to come and put everything in place on the tower to make proper connections. They'll give us time to put the new transmitter into their building at the foot of the tower.

Assembling the antenna in sub-zero weather for Fort Vermillion

Michael supplies us with boots, parkas and warm thick gloves – not the easiest to handle the radio parts but a necessity in the extreme cold. We have to measure out the thick coax cable on the ground in relation to the height of the tower. Creak, creak the cable unwinds in the cold. Then we hook up the antenna temporarily on a pole to test it first – everything works great. It's lots of fun handling aluminum in the cold but we get that all completed and the studio is prepared as well. So, we call Telus and they come up on a Friday night and stay over.

They inform us that they will not climb the tower if it's colder than -25F. or if it's snowing. We pray since it has been colder than that in recent days. That Saturday morning it's -23F and it's a go. Here we are outside again laying out the cables, antenna and parts. The Telus technician climbs up about 70 feet and mounts the bracket to hold the antenna. Then they put up the coax cable and begin to mount it in place. That cold wind nearly takes our breath away but we carry on. Because of the wind conditions, we suggest to the fellow up on the tower to use extra tie-downs for the coax cable. He gives some smart remark, yanks on the cable and pulls it right out of the fitting of the antenna. Now we have to take it all down again and repair the connections on the cable to the antenna. By mid afternoon it starts to get colder. They drop everything and say they'll have to stay overnight and do it tomorrow. These four Telus men noticed our reaction when the cable broke. We didn't rant or rave or make unkind remarks. We just said we we'd have to fix it. I was later told that these Telus employees were very impressed by our attitude of just sticking to the job to get it done. That's what it's all about.

Back in the house, we warm up and have supper. Dave's on the floor pulling all the wires back in place and resoldering the connections. We listen to the weather forecast and it doesn't look good. It's back down to -45 F that night. The next morning Dave speaks at one church and I at another. Will the Telus men wait until after lunch? Arrangements are made to take the Telus men to a restaurant for a hardy Sunday breakfast, because Dave and I are due at church. The sun is shining but still it's in the mid -30s. The Telus men tell us, "It's got to be -25 or we're going home". We ask them to stay until noon to see what happens. Meanwhile, both Dave and I ask the congregations to pray for the needed temperature. As soon as the sermon is over we both leave our respective churches without even shaking hands (rude of us) to get back to the radio site. Guess what? It's -24 and the Telus boss says, "Let's get to work." Thank you Lord. Dave and I change quickly into our warm clothes. Everything connects as planned. We turn on the transmitter and watch the meter. Everything is working first class. Our God answers prayer.

This station has carried on for years and has grown to where they now have six repeater stations. Thousands of First Nations people are hearing the Gospel in their native tongue, some using GALCOM solar-powered fix-tuned radios. One day we will see many of these brothers and sisters in glory face to face. Northern Canada Evangelical Mission has kept that ministry going with Michael still overseeing the broadcasting. What a blessing!

14

CENTRAL AMERICA, MEXICO AND SOUTH CAICOS ISLAND

"He shall call upon me, and I will answer him: I will be with him in trouble; I will deliver him, and honour him." Psalm 91:15

BELIZE

PUNTA GORDA

After many weeks of preparation, our Punta Gorda, Belize trip got off to an early 3:00 a.m. start. Tony DeWeerd from Rehoboth United Reformed Church had rounded up several others to help. We flew from Toronto to Miami then on to Belize City. We'd been praying for reasonable passage through customs since we'd heard they can exact exorbitant duties on imports of equipment. Praise the Lord, we only had to pay $87.00 on the $1,200.00 worth of goods.

Our next flight took us from Belize City to the southernmost town of Punta Gorda, home of many of the Mayan Indians. There was a small six passenger single-engine plane to carry six men, their luggage and equipment. Their weighing system for determining the acceptable load for the plane consisted of a bar snapped onto the tail of the plane cut to a specific length. When the bar touches the ground, they're at capacity. The pilot signaled for the four volunteers and myself to pile in. We never were sure if the weight of six men was figured into their calculations. John Langandoen wanted to sit in the front beside the pilot. The rest of us get strapped in ready for take off. We all pray silently and I notice the look on John's face as he watches the gauges as we speed down the runway. He told us afterwards that the torque and RPM were at maximum when we reached the last bit of runway. The plane lifts and soars just barely clearing the waves of the Caribbean Ocean. Whew! In an hour and a half we arrive in Punta Gorda with a good "thump" and unload. We realize there is no one to meet us.

A half-hour later a River of Life missionary arrives with a truck. We load up for the 45 minute drive into the jungle. We're in one of the most undeveloped areas of Belize: mostly jungle with pockets of Mayan villages scattered throughout. Our target audience is the 40 – 50,000 of these precious people who need to know about the Saviour who came to bring them eternal life. We're told that about 60% of them are Catholic in name only: this is mixed with witchcraft and voodoo practices. They are a very poor people eking out a living on small plots of ground.

We meet Dale and Barbara Sandbek, missionaries with Global Outreach who have a real heart for these people. However, they're only on a one-year assignment and must get this radio station up and running quickly. The walls of the studio have already been erected prior to our arrival. We need to put on the roof, do the electrical installation, erect the 72-foot tower with its base and three guy wires, and install and wire up the studio equipment. The Sandbek's home is situated along a small river and just the previous week there was quite a stir over an eighteen-foot alligator who had parked himself on their lawn. Fortunately, during the ten days I was there, I didn't see him.

The next morning we start on wiring the block building and preparing to pour the cement roof. Frequent hurricanes in the area dictate that the building must be solid and reinforced with rebar. While Dale, Pastor Enns, from River of Life, and I worked on the wiring, the others were busy cutting and preparing several hundred poles to support the base of the ceiling/roof. The poles are cut precisely to the height of the building and stand vertically in a grid about 14 inches apart. This is a temporary support which is then overlaid with plywood and covered with cement. For this reason, the wiring needed to be done first since it was impossible to move around inside this grid work. We also had to drive a ground rod eight feet into the ground using just an axe – we finally made it. With the wiring finished and the poles ready to be stood in place we looked forward to the next few days work.

Over those next days, the poles were put in a section at a time with wire reinforcement. An antique tractor was available to operate an old cement mixer with a power take off on the back. Then, using five gallon pails, the men worked in a chain gang to lift the cement to the roof. What heavy work! Praise God they did it. In the meantime, we had dug a square four-foot deep base for the tower by hand, gathered rocks to fill it and then poured more cement into this base with four bolts fastened to the plate that will secure the tower. Tony inscribed in the cement *"The name of the Lord is a strong tower; the righteous run to it and are safe."* Proverbs 18:10. My prayer is that the Mayan people will listen and grasp the message that comes from this tower of the good news of Jesus Christ.

The next step is to raise the 72-foot crank-up tower. But this presents a challenge. As I check the top of the tower, I see it's designed to mount a horizontal antenna, and we're using a four-bay vertical antenna. What do you do in a jungle where there are no hardware supplies? We pray!

At that point I have to make a call of nature into the bush. As I walk along, I suddenly spot an abandoned bulldozer sitting there in the bush and notice on the front of the engine a three-bladed fan, not the common four-bladed type. My mind immediately pictures bending the three blades down to form a base on which we could then mount the antenna. If we cut a piece off the shaft of an old hydraulic cylinder and weld it to the base, this would be an ideal support for the antenna. We also noticed on our drive into the area, just a few miles away, a fellow with a little welding shop. Maybe he'd weld it for us. In a short time the bracket is built and praise the Lord, eight years later that fan part is still doing the job.

The following morning we have another challenge. The unextended, telescopic tower is 24 feet long and has to be moved into place on the cement base. Its weight is close to a ton and six men cannot move it because it is lying on the ground 200 feet from the prepared tower base. Now Lord, we need your help again. Would you believe it? Just shortly after puzzling and praying about this, a van pulls in with five men who are going to work at the River of Life Ministry. We're not long asking for their assistance. They graciously agreed and using extra poles that had been cut for the roofing, we space them out through the bars of the tower to make carrying handles. With a man on each handle we lift, carry a little distance, rest, lift again and so worked our way to the tower site.

We assembled the four bay antenna, mounted the tower on the new cement base, cranked it out while it was still horizontal, attached the antenna and mounted the guy wires. While it was still in this position, we fastened down the coax cable. We then locked the tower into its extended position permanently. Now the task is to raise it on its base. At that precise moment, a man drives up with a winch on his truck: "Idea!" Would he use that to help raise the tower? He graciously agrees. Again, with the help of the five extra men, we raise the tower up about 15 feet using the long poles. Meanwhile, the guy wires are being stretched into place. The driver with the winch then pulls the tower into position. From there it was a straight forward matter to finish securing the guy wires and bolting the tower firmly into place.

Because we now had to let the cement on the roof cure for at least 10 days, we couldn't set up the transmitter and studio inside. We prepared everything possible for the hook-up and then decided to build a small shelter just outside the building to temporarily house the studio. Before long we were on the air. What a thrill it always is when you first flip that switch and suddenly hear the Gospel coming from those little solar-powered fix-tuned radios that we'd brought with us for the people. This was the first Christian radio station now sharing the wonderful message of Jesus with the Mayan Indians of this area. What an honour to be able to serve the Lord in this way! It's time to head home so we leave the remainder of the setup to Dale Sandbek who is quite capable of handling this. His wife, Barbara, a beautiful, outstanding soloist had sung at many concerts in the States. I'm sure her music, even in English, will be a blessing to the Mayan people. After eight years of broadcasting, how many of these people will we meet in heaven? God was in it, you prayed, we worked and the Holy Spirit brought in the harvest! To God be the glory!

SOUTHERN BELIZE

It was a special pleasure to have my daughter, Ruth Anne, accompany me on one of the trips to Southern Belize. Here is her story:

"From the time I was very young, I remember watching my father with pride as he helped anyone and everyone in need. People followed him because he is a true servant leader. He loves people and they know it. He leads by example and works harder than anyone I have ever met. He takes time to meet with the sick, the vulnerable and the elderly. He always has a huge smile on his face when he greets people. Most importantly, he's the same kind, gentle and loving man at home as he is in public. I saw these familiar traits of kindness, gentleness, love and leadership on this trip.

"As a child, I always loved hearing my dad's stories when he'd come home from Africa, Asia, South America and the outermost parts of the world. We didn't hear about the economics or politics; instead, we heard about the people: their families, communities, and unique personalities. We heard about the kindness that was extended to my father by the locals. They'd share their very best meals of meat and rice with him even when they had very little for their own families. And we heard about some other dinner delicacies, not so palatable, like bat, monkey, guinea pig, and so on…they made for some great oooohhhs and ahhhs at our dinner table. And I remember my dad's laughter and animated face as he'd describe characters in the stories to us, and tell us about the friends he'd made and his incredible adventures in the darkest mud huts of Africa, and on the most remote islands of the Pacific. It was with great anticipation that I embarked on my first international trip to Belize.

"My dad had been working with two contacts in Belize for months to plan the location and specifications for the new radio tower. There was already a small radio tower and station but the signal was not strong enough to reach many of the remote villages. They needed a higher tower. My dad agreed to take on the challenge and had raised funds to pay for the trip and supplies. The team members also paid their own way to help with the construction costs.

"The site was a remote section in southern Belize, a barely accessible jungle area where radio is the best way to communicate. Mostly Mayan Indians live here in primitive conditions, in rural villages, with little or no access to doctors, education or modern comforts.

"My dad assumed that we would have 'very nice accommodations' for the team when we got there as he had been there before. So, with a few work clothes packed, some sandals and running shoes, at 4 a.m. we were off. There were five of us on the team: Cali, myself, Ron, Hartley and my dad. Ron and Hartley were going to be constructing the radio tower. Cali and I did not have much experience in construction but we were willing to do whatever was needed. My dad was the team leader.

"Our flight departed at 6 a.m. and would require 2 connecting flights making the total travel time about 14 hours. When I asked about this, there was one answer – cost. Every dollar saved on travel was another dollar that could go towards radios. My dad was so committed to this he would save every cent entrusted to him so it could be used for the mission. He would bring his own sandwiches instead of buying food, take the least convenient flights often with long layovers because they were cheaper, and would give up his sleep and other comforts so that more money could go to radios. I really respected him for that. I had been on a number of fairly comfortable vacations myself and now it was my turn to travel Dad's way.

"When we finally arrived in Belize, we were tired and ready to see our 'very nice accommodations', unpack, and get a good long 8 hours of sleep. However, our mission contact, was nowhere to be found. We went outside and waited with our luggage and still no driver. We were all trying to wait patiently but to be honest, we were all very tired. I looked at my dad. I could see that he was tired but he barely showed it. He told us not to worry while he made calls to confirm our transportation. Finally, an hour and a half later the van pulled up. We loaded our luggage, got in and thought we would soon be there. However, the driver had some stops to make while he was in the city. No air conditioning, no supper again we waited in the van in stifling humidity and heat. After a 4 hour drive turned into 5 hours, night had fallen and we finally arrived at our 'very nice accommodations' at the mission base.

"Dad was told he would stay at the missionary's home a few miles from the compound, the other men were to stay in the 'guys' cabin' and Cali and I would stay in the 'girls' cabin'. Next thing I knew Cali and I were standing there with our suitcases in complete darkness. Cali was well prepared with her work boots on, long pants and flashlight in hand. I had packed rather hurriedly and was wearing sandals, light clothing, no flashlight, bug spray or anything else you might want in a jungle setting. As we started toward the girls' cabin, we were surrounded by the damp jungle air, screeches from the jungle, and bugs flying around that we couldn't see until they landed on us. I was trying to swat bugs while pulling suitcases and trying to stay on this narrow path into nothingness. Thankfully some local boys from the camp came up beside us and helped us to our cabin. It was a wooden slat cabin, up on stilts, with some makeshift stairs leading to a small wooden door.

"During our long drive, there had been some talk about scorpions, snakes, tarantulas and huge iguanas. I was trying to put this out of my mind as we entered the dark room. I noticed immediately that the wooden slats forming the walls of the cabin did not meet, so any creature the size of a lizard could easily crawl in or out. Lizards had already made their way into our little room, and were perched comfortably at various heights on the wall. They were staring at us with their little green eyes as though we had invaded their space. I guess we had.

"The washroom was over 50 feet away. As we walked there, bats flew past our heads, small snakes slithered across the grass, and we could hear the constant buzzing of bugs. After we made it to the building, and stepped through the door, I noticed the floor was moving. Was I just tired? No. the floor was moving. How strange! It was so covered in bugs, ants and spiders squishing under our feet, that it literally looked like it was moving.

"After returning to the cabin, I noticed the beds, made of plain wooden boards, had what looked like an enormous kitchen sponge on each one. We started gently poking the 'sponges' top and bottom one by one to see if anything was on or under them. Then we saw it, on one of the top bunks there was a long black dead bug. We edged closer and inspected it. There was a dead scorpion in our cabin on one of the 'sponge' beds. After disposing of it, we thankfully and miraculously went to sleep.

"So, our first day began with Hartley and Ron welding metal framing for the radio tower, while Cali and I went with my dad and some nationals to the place where the radio tower was to be erected. My dad explained that we had to find three separate spots on the ground for the guy wires for the radio tower. We determined the correct locations by walking back and forth through the fields holding ropes to get the right placement from the base. Once we found the three spots, we were going to be digging holes, inserting rebar, and then pouring concrete, so calculations had to be accurate. Finally near the end of the day we got it right and mapped out our spots. Job done! We felt so proud and were filled with such a sense of accomplishment. We happily got into the van and returned to the compound.

"We had dinner, chatted, and heard a bed time story by one of the locals about the jumping snakes indigenous to Belize, called Tommy Goff snakes. We were told that they were the only snakes that wouldn't slither away from you. They were aggressive and would jump, striking the head or neck area. So, we were told to be careful. Each night we would hear a similarly chilling bedtime story: the night screechers, tiny owl like birds that made a very loud screeching sound as soon as it got dark, the jaguar the kitchen helper told us about that had been stalking the girls' cabin. Then we heard about the three colours of eyes that could be seen through the thick jungle brush at night. green eyes would signal the common iguana, which was the size of a man but probably wouldn't jump out of a tree onto us; red eyes would indicate the largest rodent in the world, called the queen's rat (capybara); yellow eyes should compel us to run to the cabin and slam the door because it was a jaguar. This really helped with the sleeping at night.

"So, after making it through our bedtime story and a thankfully uneventful night's sleep, the next morning we woke and joined everyone for breakfast. We noticed that Hartley and Ron were smirking and smiling then erupting into full out laughter. We just had to know what was going on. After prying the story from them, they finally told us that a bat had managed to get into their cabin. After failed attempts to capture it, they turned off the light, turned on the fan and in moments a strange clonk announced the demise of the creature. After this inspiring story, we went to the construction site and loaded up rocks all day long, in the heat. We loaded wheel barrow after wheel barrow with rocks, and trekked them back and forth from the driveway across the field to each of the three sites we had previously marked out. We piled up the rocks making huge mounds for the base of the tower. Near the end of the day, the director showed up at the site. There was some discussion with the neighbor about where to put the base. He left. We returned to the compound for dinner, tired and hot but with a great sense of accomplishment after another good day of hard work. After another rousing bedtime story from the resident snake killer we went to bed.

"The next morning just as we were ready to pile in the van and head to the site for day three, the director arrived and told my dad that they didn't own all of the land where the tower was to go and that he couldn't put the tower there. Apparently, half the field belonged to the neighbor. The project came to a halt.

We were all in shock trying to process this. All eyes were on my dad. His face did not change. He looked at the director calmly and kindly and simply said, “Well what else can we help you with? The Lord may have a different plan. What else do you need? How about the radio station? Can we help your team finish the construction there?”

“The director paused and then agreed. There were locals already tasked to build the radio station but the project had been moving slowly. We could go and work alongside them if we wanted to. With a smile my dad said we’d be happy to help with the radio studio. I have never in all my years seen a better example of leadership. That was a moment when team morale could have deteriorated. My dad was amazing in that moment. He was humble, flexible and enthusiastic: qualities which are extremely important to have on a trip like this. I was so proud of him.

“I could tell Ron and Hartley were disappointed because they had worked so hard on the framing for the radio tower. They had worked late the night before to get it finished, and I recalled seeing them arrive for dinner, tired and muddy, but with such a look of accomplishment on their faces. They had done such a great job on their part of the project. I really felt badly for them. However, my dad acknowledged their hard work and reassured them that their work would not go to waste. He told all of us that everything was going to be okay and then started talking about the new project. His enthusiasm for this new project was contagious. In a few minutes we all jumped in the van, almost giddy as we drove over to the new radio station site.

“When we arrived at the location, the building appeared to be about half complete, with a few walls made of concrete blocks already constructed. Our arrival seemed to bring renewed energy to the workers already there. We carried bags of cement, rock and stone, piled up concrete blocks, and helped with the masonry work. The locals weren’t used to having men and women work together on construction sites but they welcomed us all warmly. We ended up making a lot of progress over the next few days. It was great to see the studio building taking shape.

"It seemed that all the local workers, and most of the people we met, had machetes - the tool of choice. They would disappear into the jungle and we could hear in the distance 'swish swish swish'. Branches were crackling, trees were falling and then they would reappear, and although the men were only five feet tall, they were carrying eight or nine small saplings in one arm. They were incredibly strong. We also watched them at break, as they would sit down in the shade and use the same machete to cut up their fruit. The men also used the machetes for cutting grass. Crouching down they would swish the blade back and forth to clear a path.

"This really caught my dad's eye! He wanted to try his hand at it so he asked if he could use one. The worker looked down at the machete in his hand, sort of shrugged, half smiled, and handed it to my dad. To my dad's delight, it was now his turn! He smiled, suddenly seemed full of even more energy than usual, and began to cut the grass down as we had just watched one of the men do. I noticed people gave him a healthy berth so he could practice his lawn cutting skills. They did not know what to think. For me watching…it was hilarious! After about 10 minutes, my dad got all his swinging out. With a satisfied smile on his face he was ready to hand back the tool. The owner of the machete looked grateful and pleased when my dad said a big thank you and handed the machete back. That was our entertainment of the day!

"After completing our work on the radio station, we took a short trip to visit the original smaller radio station that my dad had installed a few years prior. It was really great to see the station at work. Music flooded the barren land outside the station and I was surprised to hear them playing not only a familiar tune, but one of my favourite songs. I took a moment to look around at the primitive conditions and thought about the contrast between the thick remote jungles in southern Belize and the nice vacation resorts in northern Belize. Although it was a tough life in these areas, I was happy to know that they had the opportunity to hear these radio broadcasts and to know that God loved them through these radio broadcasts.

"After our visit to the radio station, we had heard about a distant unreached village, so we got back in the van and travelled for most of the day to try to reach this village and give them some radios. My dad, Ron and Hartley volunteered to sit on the bare metal in the enclosed back of the truck for the first half of the trip. Cali and I kept hearing laughter and it sounded like so much fun we offered to take a turn on the way back. It was actually very hot, rough and bumpy. I am not sure how my dad made it seem so inviting, but that was my dad. He could make any situation hilarious and enjoyable, no matter how unpleasant it seemed to others.

"During our drive to this distant village, we passed by a few other smaller villages and I saw some small children outside the huts. I noticed immediately that their clothes were worn and tattered. My heart went out to them. We handed out a few radios and some gifts, which the people appreciated. I was humbled by the kindness and sincerity of all those we met. They welcomed us into their homes and shook our hands warmly.

"Just as we approached the unreached village, we ran into a roadblock. The bridge was impassable. Due to the recent rainfall, the rushing river was overflowing the bridge which was submerged by over a foot of churning water. We would be in danger of being swept off the bridge and onto the jagged rocks below. We had to turn back. Disappointed that we wouldn't reach our goal, we tried to think about how we could get the radios to that village.

"Out of nowhere we saw some men approaching us. They had been walking bare foot for miles and miles to gather wood and supplies for the same village we were trying to reach. They were weighted down with supplies. One of the men, carrying a large bundle of sticks stopped to speak to us. We gave him several radios and he carried them cautiously across the bridge hanging on to the single railing. Our radios would make it to the distant village after all! As he left, we were thankful that these villagers would hear the message of love and hope, and learn about a God who was bigger than the jungles, mountains and oceans.

"As the truck turned around and we started to head back, I thought about what I saw when I entered the radio station earlier that day. The local broadcaster with a big smile and gentle eyes sat at the desk with his hands on his headphones speaking into the microphone. He was an instrument to share this message over the airwaves and into the hearts of thousands of people in remote jungle villages. Just like us, they had hopes and dreams, fears and hurts. Just as we had briefly experienced the heat of the jungles and the difficult working conditions, every day of their lives they were experiencing these difficult conditions. They needed the God of all comfort as well. And He met them there. They needed to know that people cared for them and wanted to partner with them. That is what our mission was about and that is what sending those little radios is about, to let every person know, in the furthest corners of the world, that they are loved."

SOUTH CAICOS ISLAND

February, 2012 we're lost at sea in an 18-foot boat with eleven people on board. It's pitch black in the wide open ocean: a trip that should have taken about 75 minutes has now taken over three hours with no sight of land. Our driver continues to tell us "just another half hour."

Back in 1995, we had been able to supply Pastor Alex Minott with the equipment to put up a radio station on the island of South Caicos, just off Cuba, and provided him with 1,000 fix-tuned radios. New Life radio FM 105.5 operated for over 15 years. Then in June 2010, I returned to the island to install a GALCOM Cornerstone Transmitter and send the old transmitter out for repairs. The station needed to be upgraded and the challenge was to reach all seven main islands of the Turks and Caicos Archipelago. Plans were put in place to put in a new tower, replace the antenna with a new higher gain one, and construct a building for the studio instead of using a room in Pastor Alex's home.

In early December, Tony DeWeerd from Rehoboth United Reformed Church happened to drop in to see if we knew of any radio station projects that a team from their church could be involved in. Several years earlier, they had worked on a station in Belize. I told Tony about south Caicos and before long a team of seven plus two others and myself were making preparations for the project. A 90-foot tower was to be erected with a new powerful antenna. The rebuilt transmitter would be reinstalled in a new studio building that was to be constructed.

The team left Hamilton February 6 with 23 bags of equipment, 500 radios and tools – over 1400 pounds. As we flew towards Providenciales, we prayed, "Lord let the airport accept this". In a matter of a few moments not a piece of luggage was opened and we were cleared right through with over $13,000 of equipment! We found out later that this customs lady's mother attended Pastor Alex's church in South Caicos.

Normally we would take a 20-minute flight on a small plane to South Caicos but with 11 people and all the heavy equipment, Pastor Alex had arranged two boats to take us 60 miles across the huge bay. We arrive at the dock and see two

18-foot outboard motor boats with a driver and a boy for one boat and a driver for the other boat. We put the three younger married men in one boat with a lot of the oversized equipment and baggage and eight of us with the driver and his son in the other boat. Some luggage was also in our boat along with the driver's supplies.

By 5:00 we are still getting loaded; we were supposed to leave at 3:30. I notice there are no life jackets, oars, GPS, depth finder, light, anchor or bailing can: just a 200-horsepower engine and a tank of gas. "Lord, look after us," I pray. Off we set out into the open water. We leave slightly ahead of the other boat while they finish loading. A little while later, they zoom past us and are soon out of sight.

Shortly after, we hit some shallow water and the boat is scraping sand reefs and slows down. Darkness begins to settle in and we cannot see land in any direction. I was told the trip should take an hour and fifteen minutes and we are now over three hours. The blowing wind is creating fairly hefty waves and we're getting soaked by water cascading over the bow. We bounce up and down and really start to feel the constant jolts on the wooden benches which are quite unforgiving. I am praying, "Lord, don't let us run out of gas." The driver is still saying, "Just another half hour". After approximately four hours we spot a light on the shore and eventually make land. "Thank you, Lord."

We expect to see the other boat already there since it passed us earlier but there is no sign of it. Pastor Alex, who meets us at the dock contacts the coast guard for help. "Sorry, we have no gas," he is told. "You'll have to wait until tomorrow." Our team is greatly concerned about the safety of the other three men, all with young families. "What are we going to do, Allan" they ask? First we prayed, and then Pastor Alex showed them to their accommodation to get settled in. Pastor Alex and I head down to the dock every hour until about 1:30 a.m. Finally, we agree. We just have to leave them in the Lord's hands.

Early the next morning, we're down at the dock and the driver of our boat gasses up and starts out to find them. About that time, I get a call from my son Allan, Jr. that Florrie is in the hospital with a possible heart attack. I am 2,000 miles away with a very anxious team concerned about three missing members and also, I am the only team member who knows how to install the radio station. I stand at the dock waiting for word from our boat driver who is out searching. About 8:30 a.m. we see our boat driver heading for shore with the men and equipment loaded into his boat. Praise the Lord!

Apparently, they'd hit a reef in the dark and stalled the engine which they couldn't restart. Fortunately, their boat had an anchor to keep them from drifting during the night. In the pitch black with no communication equipment, they had no way of notifying us. They had piled suitcases and bags of radios around them and tried to sleep as best they could on the floor of the boat. They had previously opened several of the suitcases to retrieve some snack foods for supper. When our boat found them, it stopped alongside and just at that moment lost the shear pin out of the propeller into the water. The only way to get back was to take the shear pin out of the disabled boat and put it in the functioning one. After dropping the men and equipment off at South Caicos, our driver went back to rescue the other boat.

It was a great reunion to get all the team back together. We started work on the tower and building. I had to make arrangements to get another engineer, Ron Marland, to fly out from California to carry on the work so I could return home. It wasn't until Wednesday that Ron flew to Providenciales and I started back to Canada. By God's timing, we had a fifteen minute meeting at the Providenciales airport as we awaited flights to discuss the project. By God's grace the team did a great job getting the tower up. At just the right time a power hoist was available to raise the steel tower into place.

A big challenge was the solid rock that comprises the island. Our team had to find a way to anchor the tower and the guy wires in the solid rock. They were told to look for small trees growing out of the rock. They managed to find little trees in all the right locations. Pulling them up by the roots, they found crevices in the rock big enough for the anchor rods to be cemented in.

We needed more cement for the building and the anchors but the local supplier had none. Pastor Alex knew just the contractor to contact and he loaned us what we needed. The team needed three more days to finish pouring the cement roof but it was really threatening rain. They prayed, and the prayer chain prayed, and the rain held off until the roof was poured and hardened enough not to be affected by it.

It was amazing that we had a radio engineer, Bill Tidwell from WAFT radio in Valdosta, Georgia, who was available to return later in June to install better studio equipment and make the radio station upgrade a reality. More people than ever are now able to hear the wonderful Gospel message.

Also, back at home, Florrie had not had a heart attack but only a slight high blood pressure incident. What a great God we serve! We thank the Lord for guidance in this much needed project. We provided radio equipment plus fix-tuned radios from GALCOM USA and Canada for 500 families to hear the message of salvation through Jesus Christ. God also allowed us to provide expertise to upgrade this station for Pastor Alex. To God be the glory for watching over us in an event we will long remember.

PANAMA

Recently a Christian missionary was travelling on a remote trail in a jungle area of Panama. He was stopped by a local villager and when the villager discovered he was a Christian he asked the missionary if he would come into the village and serve communion to the Christians there. The missionary asked if any other missionary, pastor or evangelist had been to visit them. "No," replied the man, "but there is a large number of Christians in the village". The Christian worker asked how they had heard about Jesus and the man pulled out a Galcom radio tuned to the HCJB radio station. "This is how we heard about Jesus. We decided to walk in Jesus' ways. We started a village church listening to this one little radio. Now we want someone to come and serve communion to us." God, in His amazing way, was at work in this village long before any Christian missionary arrived.

I remember Bill Bright, founder of Campus Crusade, saying to me, "God bless you Allan for the GALCOM ministry. "It's changing the world." And so it is, one life, one village at a time.

Demonstrating the equipment for Dr. Bill Bright of Campus Crusade

PARACHUTING INTO MEXICO!

Mexico has always been a challenge to reach with the Gospel, especially the many remote, almost inaccessible Indian villages strung out along the high rugged slopes of the steep Sierra Madre Range. HCJB Radio installed a series of broadcast towers along the US border beaming into Mexico, especially to the indigenous Indians there. But the only way to get in radio receivers and Bibles was to backpack them from one little Indian hut to the next, trekking over torturous mountain trails. This was exceedingly time-consuming, and dangerous.

Exciting moment to get a Galcom radio from the sky

Jerry Witt, Jr., with Wings of Mercy Mission, thought of a better way: parachute them in! He designed a little parachute made from old bed sheets. Jerry revved up his little Cessna and began flying from village to village. What amazement for these primitive people to see the little parachutes floating down into their village!

To open the little box and find a real GALCOM Go-Ye Radio! To flip the switch and hear God's Good News coming to them in their own language!

Much had been done to preserve the culture of the indigenous Indians there, such as the Huichols of Las Latus. But no one talks about the child sacrifices that still are part of their religious practice. During one GALCOM air blitz, a little radio dropped by parachute right onto the sacrificial altar! Tens of thousands of GALCOM's short wave little missionaries have gone into Mexico this way. And untold hundreds, perhaps thousands of indigenous people have accepted Christ as Saviour and turned from their animistic traditions; they've burned their fetishes, and stopped burning their babies. Although they have suffered ostracism and persecution, they have gained peace in their hearts and peace with God. This is what GALCOM is all about.

15

SOUTH PACIFIC

"Let them give glory unto the Lord, and declare his praise in the islands." Isaiah 42:12

MICRONESIA

BACK IN SEPTEMBER 2002, I WAS INVITED TO INSTALL AN AM radio station on the Island of Chuuk in Micronesia. It was decided to use an AM transmitter that James Cunningham had built (like the one we had used in Haiti) and he joined me on this project.

We flew to Guam and visited the TWR (Trans World Radio) short wave station while waiting for our flight to Chuuk. It was very impressive to see their global radio broadcasting outreach. We arrived in Chuuk at the Independent Faith Mission where we met Thomas Philip. He has a nice building on this island but very little land. AM broadcasting requires a ground plane of wires buried just below the surface that radiate out a couple of hundred feet in all directions. We eventually got the ground plane worked out with wires around the building. Then we set up the antenna and transmitter and connected everything to the studio we had installed. It was very hot, humid weather. With Jim at the transmitter site, Thomas said, "Allan, jump in the boat and we'll go out on the ocean and see how many islands are getting this signal."

Now, I'm always careful around water – even a bit nervous. Anyway, I get in the boat with my equipment to measure the signal strength and off we go. Suddenly, I am aware there are no life preservers, no anchor, no oars – just an old motor boat. "Lord watch over us," I pray silently as we bounce over the waves. We zoom ahead to one main island and we're picking up a strong signal.

It's lunch time and the natives on the island invite us for lunch. In their living room is no furniture, just some cardboard on the floor. We sit on this leaning against the wall. I look into the kitchen where a lady is squatted down cooking over a little stove. I'm not sure what I ate but I have learned to ask no questions. They brought in two coconuts and with one quick chop of the machete they cut one end off just like that and we had fresh coconut milk to drink. Have you ever tried to drink out of a coconut shell without slobbering down your shirt? Quite a challenge! The people were delighted to receive the GALCOM radios we brought and fascinated at how they worked. As we travelled back, we committed the harvest from these radios into the Lord's hands.

We arrived back safely, had a dedicatory prayer for the station, and the following day by small plane moved on to the island of Yap.

This is one of the most unusual groups of people I have ever encountered in my travels: very friendly, dark-skinned, long, black hair. Our task in this jungle area was to replace the non-functioning antenna on the tower which was located high up on a hill. I walked a short distance into the jungle and there were a number of native men hacking a canoe out of a big log with axes. I thought I'd travelled back in time. I watched for a few minutes as the men chopped away one chip at a time and then they invited me to try. I took a good whack at it and, wow, what hard wood! They certainly have perfected the technique. Their homes are made with thatched roofs and sticks for walls. These people still use stone money for currency - round stones with a hole in the middle. The bigger the stone the more valuable it is. The government now has a law to protect these coins from would-be collectors and they are not allowed to leave the island.

We had to lower the tower and the men worked hard to drop the 40-foot length with ropes. We were able to remove the old antenna and attach the new one. Back to work went these agile men, climbing like trapeze artists in the trees with ropes. They had the tower up again in no time. We repaired the transmitter, tested it and everything worked great again. Praise the Lord!

It was the end of the day and we were invited, since they had caught a huge sea turtle, to have some soup. It tasted a bit fishy but was okay. The next day, we went over to Saipan and met with FEBC (Far Eastern Broadcasting Corporation) staff at their large short wave station. I met David Creel there and that meeting developed into a working relationaship. He has since been seconded to us by FEBC to serve with GALCOM as a radio station installer. After many years with FEBC, he was anxious to return home to help his aging parents and thus was looking for another opportunity to use his skills for the Lord.

Jim headed back to the US and I had one more stop at Pohnpei, another remote island. We met with Pastor Nob and his wife, Silvia who serve with Pacific Missions Aviation. They have a great ministry on the surrounding islands to about 30,000 people, over half of whom are under the age of 25. They knew very little about radio but God was tugging at their heart strings regarding this new way to meet the spiritual needs of the many people on Pohnpei and the surrounding islands. We discussed the potential of setting up an FM station but in order to reach the people on the outlying islands, we decide to use a tropical band system. This would enable us to reach the more remote islands as well.

In 2008, Dave Casement and I went out to install the equipment. The FM station was soon up and running and they have had a great response on the island of Pohnpei with this station. Tropical band requires more engineering to fine tune the signal to the outlying islands. These are some of the challenges we face in broadcasting as the signal was skipping some of the islands but being picked up as far away as Japan. Dave Casement and David Creel over several years battled this challenge and now, praise the Lord, the signal is reaching most of the islands in the area. Since then, Dave Casement has returned to install two FM transmitters to provide full coverage for the island of Pohnpei.

We thank the Lord for the open door the Lord has given us on many of these remote islands where God is building His Church as the Gospel goes out over the airwaves.

INDONESIA

Does God Answer Prayer? Our trip into West Timor in Indonesia provides plenty of affirmation to that question. Back in May 2010, I travelled with a team of seven people to install an FM Christian Radio Station in Atambua, West Timor. Pastor Paulus, of Mercy Indonesia, had helped thousands of orphaned children especially after the huge tsunami that swept across that area in 2004. Daily feeding, clothing, housing and educating thousands of children presented a huge challenge. Consequently, the government wanted to reward him in some way. They asked him what he would like. "I would like 100 FM licenses for Christian radio in many of the unreached communities of our country," replied Pastor Paulus. In recognition of all that he had done, the licenses were granted.

Now, here I was with a team to help install one of these stations. After 27 hours of travel we eventually reached Denpasar.

We approached customs with the transmitter, radios, equipment and tools – over $15,000 worth. They were about to levy a huge customs fee. We requested to speak to their supervisor. As they went to find him, we huddled in a group and prayed for the Lord's help. After a long wait, the man returned and said, "Just pay 10% of the value and you can go through". We praised God for answered prayer.

Two of the ladies on the team were staying in Denpasar to help Pastor Paulus with the children and the next day the remaining five of us were to travel by air to the island of West Timor to Kupang. David Creel, one of our Galcom engineers, Steve Moore, a volunteer and former radio production worker at GALCOM , Don McLaughlin of Bible Voice Broadcasting and Wayne Hill, a solar panel specialists and myself arrived at the airport about mid-morning. The airline was prepared to charge us an excessive amount to carry our extra luggage on the plane. Again we prayed and Pastor Ady, of Mercy Indonesia, was able to get the fee reduced significantly. We always pray for safety and again the Lord watched over us.

We landed and were awaiting our luggage when the pilot of the flight came up to me and said, "Are you Allan?" I was surprised he knew my name and told him I was. He said, "One of the cleaning ladies found this on your seat and gave it to me". He handed me my passport. I was so relieved. First, I didn't even know it had fallen out of my pocket and secondly, I was surprised that it was returned and not sold on the black market. That's the first, and last, time I ever lost my passport. Thank you Lord.

As we loaded our equipment onto the truck, we noticed there were no guy wires. Whether they had been left behind in Canada or were deposited somewhere in Indonesia we didn't know. We only knew we needed them to complete the station. We had already planned to purchase a heavy duty truck battery for the solar system to operate the radio station we were installing. Now guy wire was added to the list. Again, we prayed for the Lord's guidance. We drove down into the main business section and found a hardware store packed with every imaginable piece of equipment. We found the guy wire we needed back in the corner of the store. While I purchased the wire, two others went to search for a battery and found what they needed just a few doors down the street. Again the Lord answered prayer.

We carefully packed all our luggage on the truck and sent the driver and his assistant on to Atambua. The five of us would follow shortly after in a small car we rented. Seven hours of roller coaster riding and bouncing, starting on a paved highway and ending up on a dirt trail, we arrived in the jungle of Atambu. Thank you Lord, once again, for safe travels.

The next morning we trekked half an hour out to the radio station site. The government had built a small cement building among the refugees in that area to accommodate the leaders and which they planned to use for the radio station equipment. Just then, the truck pulled in and we were horrified at the sight. The truck driver had picked up several passengers along the way. They had shifted our carefully placed luggage and laid the glass solar panels on top to serve as a seat. For many miles five men had travelled bouncing over the rugged terrain seated on the glass panels. I saw the distressed look on Wayne's face as he took in the scene. We unloaded the equipment, eventually tested the solar panels and amazingly they were not damaged.

Our challenge here was to install a 45-Watt GALCOM Cornerstone transmitter, erect a 50-foot tower and antenna, install and connect up the three solar panels on the roof of the building to charge the battery, and to set up a wind generator on a 40-foot tower for additional power. We had four days to complete this. We all knew our specific tasks, so after a word of prayer, we set to work. You must remember we were in a remote community at this point. While drilling the holes through the tower pipe and antenna shaft to clamp them together, one of the holes slipped off centre. We needed a round file in order to correct this: otherwise the antenna would be on an angle, putting extra stress on the joint. Again we prayed for this. One of the nationals happened to have a file and brought it to us: it was triangular but we were still able to use it to get the job done. The following day, we were mounting the antennas and one of the threads got damaged. Now we needed a triangular file with an edge to repair the damage. Once again, the need had already been supplied.

While one of the Indonesian workers was attaching the antenna to the tower, he accidentally dropped one of the special nuts from about 30 feet up in the air. It fell into the sand below and was nowhere to be seen. We prayed, then got down on our hands and knees carefully searching. Praise the Lord, we were able to find it – another answer to prayer.

Both the 50-foot tower for the antenna and the 40-foot tower for the wind generator had to be sunk into three foot holes and secured with concrete. In the morning, just as we began digging, the wind picked up and a tropical storm threatened to move in. We needed time for the concrete to set and heavy rain would prohibit that. Again we prayed that the Lord would hold back the rain. We completed the job around 11:00 a.m. and spent the rest of the day wiring up the solar panels and transmitter and the studio for lights. Near the end of the day, the clouds opened up and we were drenched but at least the concrete had had time to cure sufficiently. Praise the Lord.

The next morning, travelling back to the site, we got about half way and the little creek that we normally drive through had now become a good sized stream. The vehicle slipped and slid in the muck so we parked the vehicle and walked in the remaining mile. By now our shoes were sopping wet and caked with clay. The tower for the antenna which we had cemented into place the previous day, had a telescopic extension that now had to be raised the remaining 20 feet. The antenna had already been attached to that top section which was nested inside the main part of the tower. We prayed for safety, strength and agility.

One Indonesian, Ady, climbed the base section of the tower which was equipped with small metal steps like a telephone pole. Once he reached the top, he would have to wrap his legs around the base and then hand over hand pull up the inside section of the tower. It took a great deal of strength to accomplish that. Another man, at the same time, had climbed up behind him with the nut and bolt that would lock the upper section in place. The challenge was to support all that weight while making sure the holes lined up. The second man would then reach up and quickly insert the bolt and tighten the nut. Again, the Lord enabled them to do this in safety and we were ready to fasten the guy wires in place to support it.

Once the guy wires were secured, Ady had to climb back up the base pole, shimmy up the extension and tie down the co-ax cable every two feet or so to keep it secured against the tower. Otherwise, it would soon be damaged if the wire were allowed to flop around in the prevailing winds. Again, God kept him safe.

As we ended our four-day project, we were able to flip on the switch and everything worked beautifully. After a prayer dedicating the radio station to the Lord's use, we distributed little GALCOM radios to the villagers, most of whom were refugees from East Timor. How rewarding to know that these people would now be able to hear the Gospel message daily and be discipled through His Word.

One interesting aspect of this project began months earlier in Ottawa. I was there for a conference and was speaking to one of the Life Groups (small Bible study and prayer groups) from Metropolitan Bible Church. I had known a number of these people for many years and I shared how a team had installed two new radio stations and provided equipment and training for local technicians to install five more stations in Indonesia. I mentioned that the only thing needed to install the final 51 stations was funding. It was apparent that they were interested in helping. A thought occurred to me: wouldn't it be interesting to set aside any unexpected money toward the project. Perhaps even their small group could raise the $5,000 needed for the installation of one of these stations. The idea caught on and they enthusiastically accepted the challenge. They decided to pray and trust the Lord to send in unexpected money.

In early November, I returned to the "Met" to participate in their Annual Missions Conference weekend with Florrie. While there, I connected again with this Life Group. They excitedly shared stories of ways they had been blessed with "unexpected money". Several received HST refunds that were unexpected. One member bought a coat at a garage sale for $2 - when she got home she found $41 in the coat pocket. Another volunteered to do some work for a widow and a neighbour and received an unexpected $150. Another couple had set aside a sum of money for repairs to their home but then the work was taken care of by one of their boarders at no cost. Still Another couple had decided on a price to ask for their boat, but then decided to ask for $300 more - the boat sold for the full asking price. A Life Group garage sale brought in more funds, as well as items sold that had been stored for years in someone's basement. Over a few short months, this Life Group joyfully accumulated a total of $5,757! God had abundantly blessed each one of them and in turn strengthened their faith. I have since shared this story with other groups and recently a small group in Cambridge, Ontario has agreed to take up the challenge of trusting God to provide "unexpected money" so that they too could experience the joy of giving and be partners in bringing the gospel to a needy world.

Ever since we founded the ministry almost 25 years ago, we have always made prayer a priority. God's Word says in

I Thessalonians 5:17, "Pray without ceasing". Our goal is to continue to move ahead by prayer. Many might say that these things are just coincidences, but in my experience, these coincidences only occur when we pray.

16

THE GALCOM RADIO EFFECT

"For I am not ashamed of the gospel of Christ: for it is the power of God unto salvation to every one that believeth..." Romans 1:16

THERE IS NO POSSIBLE WAY TO SHARE THE HUNDREDS AND hundreds of stories about the effect of GALCOM radios. We share just a sampling here that could be multiplied a thousand times over.

IRAQ

In Iraq a 17-year-old married Muslim woman accepted the Lord through radio, and because of that her husband put her in prison with her infant son. That month she had lost her mother and only brother in a car accident. The local church stood by her in prayer. In court the judge asked her if she had become a Christian: she acknowledge that she had come to Christ. The judge granted her a divorce and allowed her to keep her child. This is almost unheard of in such cases.

Zewar, a Christian taxi driver living in Zakho in northwest Iraq, made a practice of giving GALCOM radios to customers. One customer demanded that he renounce his faith and when

Zewar refused he shot him 28 times killing him. This is the price many people pay for trusting in Christ. There have been more martyrs for Christ in the last hundred years than in all of previous history.

PUERTO RICO

Janet Lattrell has a prison ministry to as many as 6,000 prisoners in Puerto Rico. One of the prisoners named "Big John" was very hostile toward everyone but Janet persuaded him to accept a GALCOM radio. He later testified that as he listened to Christian radio he learned how his burdens could be lifted if he received the Lord. He did repent and yielded his life to Christ, then asked Janet to start a Bible study in his cell block. He said that the very day she gave him the GALCOM radio he was planning to commit suicide. The Gospel message through this one radio saved his life, his soul and lifted his burdens. Now there is a Bible study in "Big John's" cell block with many souls being brought into God's kingdom all because of one Galcom radio proclaiming the Good News.

PHILIPPINES

We have sent thousands of GALCOM radios into the Philippines to assist with various ministries there. Many have gone to the Blind Mission and Randy Weisser working with them related this story: He said that thousands of people who walk in physical darkness have seen the light of the Gospel through listening to the little GALCOM solar-powered fix-tuned radios – oh, how they value them. One elderly blind man was given a radio which he hung around his neck during the day using the antenna. At night he would tie it to his wrist so that no one could steal it.

For the blind in the Philippines radio is a major link

It was, quite literally, his only earthly possession and was very dear to him. Daily he would listen to the broadcasts that brought him spiritual light and hope.

HAITI

What does it cost to be a Christian? For many years we've had the privilege of partnering with Radio 4VEH in Cap Hatien, Haiti. David Bustin, grandson of the founder passed on this story: Yon Yon was a former witchdoctor. He was nearly beaten to death. Then he had one eye gouged out and was dragged through a fire. This was on Christmas Eve. What was his crime? He had simply indicated to his voodoo followers at a ritual gathering that he wanted to get one of the GALCOM fix-tuned radios he had heard about so he could learn about Jesus. Fortunately, some people from 4VEH heard about Yon Yon and came to his aid. They cared for his medical needs, and after he had regained consciousness, nursed him back to health. He soon came to saving faith in Jesus. This is amazing considering all he had been through just showing interest in knowing about Jesus.

Fortunately, 4VEH personal helped him again by sending him to the coast to get a glass eye. He returned with a shiny eye and a radiant face joking that it is cheaper to buy an eye in Haiti than a big bag of rice. Yon Yon has been faithfully witnessing to his community in southwest Haiti, demonstrating a totally changed life through the power of Christ even in the face of a hostile voodoo community. I wonder how we would respond under the same circumstances. He still loves his little GALCOM radio.

17

SPECIAL TRIPS AND CONFERENCES

"... for thou wast slain, and hast redeemed us to God by thy blood out of every kindred, and tongue, and people, and nation.' Revelation 5:9

BILLY GRAHAM CRUSADE

I REMEMBER ONE OF OUR EARLY REQUESTS IN 1995 WAS FOR 3,500 radios for simultaneous translation at the Billy Graham Crusade being held in Toronto that June. We were in full swing and for the Crusade, we had built special transmitters for speech translation and had to have eight units ready for use at the Air Canada Centre. Everyone worked hard. It took many hours to wire up the required units for the Crusade but they all worked beautifully along with the 3,500 special radio receivers. We found out later that of the people using the translation system over 900 of them had made commitments to Christ. Praise the Lord! I lost count of the number of these special transmitters and accompanying radio receivers that we've shipped all over the world.

GLOBAL CONSULTATION ON WORLD EVANGELISM, SEOUL, KOREA

One of the very special conferences we were invited to attend was in May 1995 at the Global Consultation on World Evangelism in Seoul, Korea. Over 4,600 Christian leaders from 217 countries were gathered by special invitation to attend. We were invited for a two-fold purpose. First to attend the conference as a participant and secondly, to set up a translation system on 26 different frequencies and in six languages spread over a stadium that held 80,000 people. Galcom had been requested to supply just over a thousand radios to accommodate the visiting language groups. It took me a day and a half almost non-stop to get everything set up with the assistance of several helpers.

They opened the conference especially to reach university age students and over 82,000 showed up packing the place to above capacity. We brought 1063 radios with us, made in Israel at that time, and they had so many people needing radios that they had to instruct the people to buy transistor radios with earbuds or earphones to meet the need. Thousands of radios, besides the ones we provided, were used for that conference and all the transmitters performed beautifully throughout the entire 10 days. What was amazing was, one night they held a prayer vigil with over 80,000 people present. It was pouring rain but not one person, to our knowledge, left the place. They prayed out loud all at the same time for about two hours. What a moving experience. Louis Palau really appreciated our part in this conference and I wonder even to this day where all these 82,000 Korean students are this many years later. One day we will find out. This was one of the most moving conferences that I have ever attended.

GERMANY/ISRAEL/UK

In 1993, I was invited to speak as a guest at the European Religious Broadcasters Conference in Stuttgart, Germany. They would look after my travel and accommodation. A gathering of several hundred broadcasters and workers from all over Europe would be attending and I had been asked to make a presentation on the use of low-powered broadcasting and technology along with our new solar-powered fix-tuned radios.

When Ken Crowell heard I was going to be in Europe, he invited me down to Israel to meet with them, do some planning and see the new setup of radio production there. He would make all the arrangements for travel and accommodation. Also, they were having the dedication service for their new Galtronics plant in Livingston, Scotland and he wanted me to attend that as well on the return home.

As I prepared to leave I asked Florrie how much money we had for travelling. She scraped together $100.00 U.S. so off I went. This was prior to the extensive use of credit cards, especially overseas. I spoke in Germany and enjoyed a great time there making significant contacts. By now my funds were quite depleted after looking after some travel and meals.

I flew to Israel and met with Ken and his crew in Tiberias after which I was invited to leave a day earlier in order to meet with a group of believers in Jerusalem. A bed-and-breakfast arrangement had been made for me there and I was to fly out the following morning. I finished my presentation of the Galcom ministry to the people in Jerusalem and settled down for a good night's rest. In the morning, I handed in my key, said thank you and was ready to take off for the airport. The agent at the desk called me back and said, "You have to pay for your room". I told him it was supposed to have been looked after already but he insisted I pay. The cost was $35.50 in cash. I pulled out what was left in my pocket - $36.00 - and paid the receptionist. I had only fifty cents left to tip the cab driver for the lift to Tel Aviv. He must have thought I was quite stingy.

Well, on the six hour flight to London, I began to pray, "Lord, I need your help. I have no cash left and when I reach Heathrow Airport, I have to make my way to the WEC guest house in London, get transportation to Livingston, Scotland and after surveying the new plant there, return to the airport. I may even need to buy a few meals along the way".

As it happened, I was one of the last to deplane. Walking down the gangway, I noticed something lying on the floor. I looked down and could hardly believe what I saw – money. I picked it up and there were two American hundred dollar bills rolled together. I spoke to the passport clerk to page the person who had lost the money but he said he was unable to do that and I should speak to someone at the information desk. I proceeded to the information desk and it was closed until 8 a.m. the next morning. Since I had to leave the airport immediately and would not be returning for several days this left me in a quandary. A security guard nearby heard my story, shrugged his shoulders and said, "Finders – keepers, losers – weepers!" All I could say was, "Thank you, Lord." I actually returned home with more money than when I had left!

LOGOS

While I was still with Gospel Recordings, the idea for a Telephone Communication Centre came to me as I met with Fraser Churchill from OM (Operation Mobilization). This unit had eight phones spaced around a central core with push buttons to operate the phones. About 14 languages had been recorded with a short Gospel message in any of the selected languages. The unit was used in Halifax at a large missions conference as well as several other conferences and churches. Finally, it was placed on the Logos ship (OM) that travelled around the world to share the Gospel in various ports of call (over 250) and the surrounding areas. Unfortunately, that ship ran aground on rocks off Tierra del Fuego, Chile in 1990 and the Telephone Communication Centre sank with it.

SUMMER OLYMPICS: ALTLANTA, GEORGIA, 1996

Sometime later, Harold Kent mentioned to me that we should use the portable telephone centre at the YWAM (Youth With A Mission) booth at the Summer Olympic games to be held in Atlanta, Georgia, since there are many languages represented at these gatherings. Harold provided the funds and we developed a four-sided Language Communication Centre. Florrie prepared a "universally acceptable" two to three minute Gospel message. Our daughter, Betty Lynne, along with Susan Timmerman, worked hard at contacting people from all the 50 language groups in Toronto and Southern Ontario that we had selected. That was a tremendous job as Betty Lynne travelled all over southern Ontario to find translators/speakers for the various languages.

Tom Kerber, our resident engineer at the time, assisted in the assembly of this unit which also allowed for a print out of the message in the various languages. A large rotating globe with the banner, "GOD HAS A MESSAGE FOR YOU: STOP AND LISTEN" was suspended above the unit to draw people to it. That trip was a special treat because I got to take three of my daughters to help at the booth: Ruth Anne, Shari Lou and Loralee. What an experience that was to see people from all different countries coming to the booth and listening to the full two to three minute taped message.

The Language Communication Centre reached people of many languages at the Olympics in Atlanta, Georgia

They were so thrilled to hear something in their own language. One day, I believe we will meet some of these young people in heaven who listened to these messages as it ignited a hunger in their hearts to know the true God.

The first time the system was used in Toronto for a large missions conference, Tom and I were setting it up but it just wouldn't work—the conference was about to begin. I simply said: "Tom, let's pray." Right after the prayer, Tom tried a phone and it was working.

Thank You, Lord!

THE METROPOLITAN BIBLE CHURCH, OTTAWA

This is my home church and the church where my wife, Florrie, served for a number of years as Director of Christian

Education. When I married her and took her away from the church to attend Bible College, they didn't hold that against us. This church has had a great impact for missions over the years. When we started out in missions with Gospel Recordings, now GRN (Global Recordings Network), in 1982, they took on our support as their missionaries. I remember demonstrating the Card Talk record player to the people there and they were fascinated by it. It held about three minutes of material. Now, almost 30 years later, they are fascinated with the Solar Audio Bible that can hold over 800 hours of material. These Solar Audio Bibles are being produced in numerous languages and used in many countries around the world. With so many non-readers in the unreached areas of the world, this is the only way many will hear God's Word. The encouragement, prayers, letters and support of this special group of people at the "Met" has been invaluable to us and to the ministry. Their sound teaching and vision for outreach over the years certainly gave me a solid foundation in my developing years as a believer. Florrie and I are so thankful for the way they have stood with us over three decades of ministry. Several other churches have faithfully stood by us all these years including Hill Park Bible Church, Faith Gospel Church, Winona Gospel Church, Calvary Bible Church in Smiths Falls, West Highland Baptist Church and Community Bible Church. Thanks to all of you who have such an interest and investment in missions.

THE NATIONAL RELIEGIOUS BROADCASTERS CONFERENCE (NRB)

We have made a practice of attending this conference annually in order to connect with many who are involved in Christian Radio and to establish partnerships with a number of these ministries. Usually there are 4,000 to 5,000 in attendance.

Dr. Ron Harris presents the International Achievement Award to Allan McGuirl

In 2010, to my surprise, I was awarded "The International Achievement Award". How much of this recognition belongs to those who have faithfully prayed for the Galcom ministry and for Florrie and me all these years! We have been witnesses to how the Lord has opened up area after area to Christian broadcasting when it was deemed humanly impossible: countries such as Indonesia, Albania, Greenland and Iraq to name a few. To God belongs all the glory. We are now into every inhabited continent in over 140 countries with over a million solar-powered fix-tuned radios having been distributed.

Allan demonstrates the fix-tuned radio, projectors, solar P.A.s and transmitters for David Mainse

I have had the opportunity to share the Galcom ministry through another NRB member, CTV on its 100 Huntley St. program. David Mainse and his staff have always been very supportive of us.

COICOM CONFERENCE (CONFEDERACIÓN IBEROAMERICANA DE COMUNICADORES, MEDIOS MASIVOS)

This is a conference that started in Latin America over twenty years ago. It's held each year in one of the participating countries for several thousand delegates who are involved in broadcasting the Gospel in one way or another. Almost every year,

GALCOM has been there. Our booth is among the busiest displays since many of these people do not get the opportunity to see some of the equipment being developed in radio technology that could be used in their particular setting. Contacts through this conference have enabled us to reach into every country in Latin American with radios and/or radio stations. With the Lord's guidance and enablement the spiritual bondage is giving way to freedom. As God's Word says, *"You shall know the truth and the truth will set you free."* Praise the Lord!

One of the COICOM Conferences was held in Chili. After attending and discussing ministry for radio outreach in several countries, I prepared to return home. At the airport, I proceed through security and the normal events of emptying my pockets, removing my belt and shoes and heading through the scanner. I'm preparing to pick up my belongings on the other side when two security guards approach me and in broken English demand to see my passport. "Why did you come to Chili." they want to know? "You have committed a criminal offense by attempting to bring a banned article onto the plane." Puzzled and a little intimidated, I have no idea what they're talking about. Then they proceed to take one of my shoes and start ripping off the sole. Their efforts reveal a rusty broken razor blade and they accuse me of trying to smuggle it on board. A quick prayer later, I try to explain to them that it must have somehow been left in the shoe during the manufacturing process. An older security guard moves in and there's an extended conversation in Spanish. He turns to me and waves me on. I lose no time pulling on my shoe, floppy sole and all, and heading for the plane.

During these COICOM Conferences, what really has impressed me is to see so many men and women whom God is using in some of the most remote places: people who are faithfully serving the Lord often with very little. I've lost count of the number of radio stations the Lord has enabled us to help with whether a complete installation or giving advice on getting started. So many dedicated and godly men and women are lovingly serving under extremely difficult conditions in out of the way places. They will hear, "*Well done, my good and faithful servant*" when they reach their heavenly home (in Spanish of course).

It was very humbling and surprising at the conference in Guatamala in 2005 to be chosen for the annual "COICOM Honour Award" for assisting so many radio outreach personnel in Latin America over the years. All glory and honour really belong to our Lord for what He has accomplished.

PEOPLES CHURCH MISSIONS CONFERENCE, TORONTO

What a great church with a great heart for missions! Year after year we would attend and praise God for their faithful support of the radio outreach ministry. Just after Galcom got started, I was invited to attend their fabulous missions conference. I remember that for many years afterwards, they sold Galcom T-shirts at the conference to send literally thousands of radios around the world. What an impact! We were also able to set up the Language Communication Centre in this church where many nationalities are represented. I don't think Dr. Paul Smith knew what to do with me when I pulled so many radios out of my pockets.

Later, Dr. John Hull was thrilled with the ministry and always gave fantastic support. And I still remember Dr. Charles Price watching me pull 20 radios out of my pockets for our 20th anniversary in 2009 at their missions conference. Then I pulled a parachute out of a smaller pocket and he modeled it on his head. It was a humorous way to share the message that these little solar-powered fix-tuned radios are even parachuted into remote areas so people can hear "*the Gospel of Christ which is the power of God for salvation to everyone who believes"*. We humbly thank Peoples Church for the way they have supported the ministry of Galcom and have prayed with us. Mrs. Jackson and her team have been regular prayer warriors. As country after country has opened to the Gospel through these radios and radio stations, we are reminded that it is the prayers of His people that so powerfully accompany our every effort.

AMSTERDAM 2000

I was invited to this conference July 29th to August 6th, 2000. Well over 10,000 delegates were present from around the world. At our display, countless visitors talked with me and carried away pieces of information on radio outreach. This led to many new open doors into Africa, Latin America and Eastern Europe. Praise the Lord!

TWENTIETH ANNIVERSARY CONFERENCE, AUGUST 2009

This was a very special celebration for Galcom. It ushered in a change in leadership and a time to reflect on all that God has done. It was also a time to recognize two other very godly men, Harold Kent and Ken Crowell for their dedication and commitment to the vision of flooding the world with solar-powered fix-tuned radios to reach the unreached for Christ. Yes, they spoke of Florrie and myself that night, but you know it is my precious wife of whom I am so proud, who has stood with me all of these years. She has held the fort caring for our five children when I have been off to some remote corner of the globe, perhaps in a jungle working on a radio station. On many occasions we had unexpected visitors for meals or staying overnight. Many times certain reports had to get out especially in the early years when she was secretary, accountant, mission representative and publisher of newsletters and promotional literature. On top of that, she had to try to reign me in to slow down a bit and get some rest. I say, thank you, Florrie for being the godly woman you are. Without you I don't think we would be where we are today.

At the banquet I would have liked to have said more, but I was overwhelmed by the number of people present, the books that were made up of the many countries where we have worked, and the heartfelt tributes that were given. We were so grateful for two items in particular. Andrea McGuirl had put together a beautiful book of many of the trips that I had taken, and Tim Whitehead had made up a special model gold radio as a memento of the occasion. Our desire is to continue to move ahead and to reach souls for Christ under the leadership of our new executive director, Tim Whitehead. What a great God we serve!

18

FIVE RADIOS

"In the morning sow thy seed, and in the evening withhold not thine hand: for thou knowest not whether shall prosper, either this or that, or whether they both shall be alike good." Ecclesiastes 11:6

FROM THE VERY BEGINNING, WE BELIEVE GOD HAS ENABLED us to make durable, efficient, effective and high quality radios. In many places and under the severest of conditions these radios have continued to perform far beyond our expectations. The following stories are just a few examples of that.

FIREPROOF?

I remember the place so well, Cape Town, South Africa, right on the southern tip of the African continent. There's a walkway leading right down by the rocks at its very tip – beyond this is Antarctica. I felt the steady breeze and the penetrating sun beating down on me. To my left was the Indian Ocean and on my right the Atlantic, what a spectacular sight. Behind me lay the sprawling city of Cape Town. Located in that city is Cape Community FM Christian radio station. Into the studio one day came an African woman from "Shanty Town" – a section of Cape Town. "I have to tell you a story," she exclaimed, and produced from her bag one of our blue FM solar-powered fix-tuned radios. It was blackened and dirty and the label was almost illegible. It was one of the 1,000 radios donated to the radio station to distribute to the most needy families. Three months prior, this woman had become the recipient of one of these radios as they were placed in the shanty town area known as Squatter's Camp. Homes were made of cardboard, tin, plastic and whatever else could be scrounged in the vicinity. Listening intently over the following days, she had come to faith in Christ and was blessed to hear the radio programs day after day. One day, fire swept through the camp, destroying everything in sight including this woman's home. After the embers had cooled, she made her way back to the location of her shack. Kicking through the rubble she discovered her precious radio. She brushed it off, turned it on , and to her delight, it still worked. Although, she had lost everything else, her most precious possession had been spared. Praise the Lord!

AIRBORNE RADIO

We had provided Brother Tani, in Korce, Albania, with 1,000 radios to distribute among the poorer people of the community. Many of these were Muslim. Just at that time some radicals came to attack the station. Enraged, one of them grabbed a radio and using the antenna wire as a sling, flung it at Tani with all his might. It sailed way over Tani's head and landed up on a steep part of the roof of a nearby home which was almost inaccessible. God had His hand on this radio: when it landed with a thud, it turned on and even the solar panel was facing up. Soon the whole neighbourhood could hear the Gospel messages as it continued to play for months. We know from the book of Isaiah, that His word will not return to Him void. Praise God!

WATERPROOF

Our radios were desperately needed in some of the destitute areas of the Philippines. This particular area was subject to flooding and sure enough, one day, the homes in the area were inundated with water. The pastor and his family had to flee to higher ground for three days while the water subsided. Returning to their home, they began the messy and tedious job of cleaning out all the mud and sludge left behind by the receding water. In the process, he came across his little blue Galcom radio. He carefully cleaned off the muck and set it out to dry along with other things he was able to retrieve. As he returned to the house and continued the cleanup, he was suddenly aware of music drifting into his home. On investigation, he found the Galcom radio had dried out and had started to play again. How excited he was that he could still here the message of God's Word in spite of all the other loss.

DEMOLITION RADIO

Pastor Florin Pindic-Blaj and his wife, Lidiana, were always overjoyed to receive shipments of Galcom radios. Over the years, they were provided with over 12,000 radios for Moldova and Romania where they served. They loved to blitz the cement apartment buildings in Moldova erected by the previous government which housed thousands of needy families. One mother was excited to receive her radio but knew she had to be cautious. Although she enjoyed the radio programs, she knew her husband would not approve. She was careful to only bring it out when he was off to work. One day, he came home unexpectedly because he was not feeling well. He discovered the radio and in a fit of anger tossed it out the window. The radio fell the four stories down to the cement below shattering the case into a number of pieces. She was later able to rush down the four flights of stairs and carefully retrieve the pieces. The next day, she visited Pastor Florin's office and with tears showed him the radio. The technician on hand took the pieces, glued them all together and turned the radio on. It worked perfectly! From then on she was extremely careful to guard her precious radio. God's Word cannot be stifled!

NO RETIREMENT

South Caicos Island, due to the numerous tropical storms and the devastation they have created, is the poorest of the Turks & Caicos Islands in the Caribbean. Pastor Alex Minott and his wife Sharon have served there for many years. Back in 1995 we helped them set up a 10 Watt Christian radio station and provided them with 500 solar-powered radios. In 2011, we took a team of workers from Rehoboth Christian Church, Hamilton to upgrade the station. At that time, Pastor Alex took me aside and demonstrated how his little Galcom radio was still working after 16 years of use. The label was worn and the radio was showing its age but it still worked as well as ever.

These few accounts demonstrate our beginning vision to ***"provide durable technical equipment for communicating the Gospel worldwide."*** Only the Lord knows how many similar stories could be told all around the world because He chose to initiate this work of GALCOM. He is the founder, provider and sustainer of this ministry. For all that is accomplished, all glory, praise and honour belong to Him.

19

NORTH TO GREENLAND

"Delight thyself also in the Lord: and he shall give thee the desires of thine heart. Commit thy way unto the Lord; trust also in him; and he shall bring it to pass." Psalm 37:4,5

AS THE YEAR 2000 BEGAN, WE WERE LOOKING AT THE COUNtries that were still unreached by Christian radio and seeking the Lord's direction as to where He would lead us next. Greenland soon became a burden on my heart. I phoned Errol Martens of YWAM (Youth With A Mission) whom I heard had spent a good number of years in ministry there. I had met him previously at Mission Fest Vancouver. He was thrilled with the idea and explained how desperately they needed Gospel broadcasting but he was not able to do much at the time because of his work load. We agreed, however, to pray for the Lord's guidance. I learned a lot more about Greenland from our visit and Errol understood more about Galcom and referred us to Pastor John Neilsen in Nuuk. We decided to get people praying for Greenland. Over the next eight years, we found out that the Department of Communication in Greenland was not really interested in helping us get a license.

John invited me to attend their June Conference in 2005 when many of the nationals that they work with would gather together for worship, fellowship and planning for the coming year. I remember flying first to Denmark which is their mother country. Twenty hours later I arrived in the west coast community of Manitsoq. The Greenland Free Church Conference was just starting that day and what beautiful, warm people they are. The terrain is very different and extremely rocky. Situated well north of the tree line, we were only a 3 ½ hour flight from the North Pole. Greenland is approximately 2,670 km. long from north to south and about 1,290 km. wide.

The town of Manitsoq is made up of hills and valleys with most of the homes built on hilltops close to the ocean. Where I stayed, I had to climb up 101 steps to get to the door. One funny situation occurred. I was scheduled to speak at the church that afternoon so navigated the 101 steps down to the lower level. Then it dawned on me that I had left my notes back in my room. Back up the 101 steps I trotted to retrieve them and back down to the church service. Guess I needed the exercise. Toward the end of June at 11:00 p.m. the sun is still shining which is quite an experience. At that time of year there is about 22-23 hours of daylight. They often use heavy black blinds on the windows to try to block out as much light as possible for sleeping: for me, the pillow did the job.

At the conference, we heard some precious testimonies of what God is doing and even of revival breaking out in a couple of areas which is encouraging. I was asked to take part of the morning session and to share the Galcom radio ministry. I had a very positive response and throughout the balance of the conference, with the help of interpreters, I was encouraged by the good questions that were asked. They said they would form a committee of several churches and discuss the possibilities as the Lord led. There were times of prayer and it puts us to shame because when they pray, all get involved pouring their hearts out to God. I was very encouraged. The evening before I left, I was staying with a family on the edge of the ocean. That evening about seven big whales came into the bay area to feed and what a display they put on. I wish I had had a video camera or even my digital camera. They were diving and their tails would wave high up in the air. Then others seemed to bump into one another grabbing for food and diving again. Thank you Lord for such a treat!

Also while there, they had a special celebration at the waterfront. Everyone gathered as a technical crew brought in the cable for internet into the country. The large ship was anchored off shore while a smaller boat dragged the cable with a kayak leading the way. The cable between Canada and Greenland would now allow the people to connect to the internet for the first time.

Spending time with the people and church leaders, I found out that of approximately 64,000 people in Greenland, less that 2% are really aware of the need of having Jesus Christ in their lives. They have their challenges with alcohol abuse, child abuse, immorality and depression. What a place for the Gospel of Jesus Christ! It was a pull on my heart as I was leaving to learn that Greenland has one of the highest suicide rates among young people in the world. I prayed, "Lord help us to get Christian radio here soon to reach these young lives for Christ."

Shortly after the conference, the Greenland Committee met and became very active with Pastor John Neilsen giving leadership in the capital city of Nuuk. During the next couple of years enquiries were made at the Department of Communications in Nuuk to get a license but each year a license was denied. Then in 2008, two women were placed in charge of the department and when we again applied they responded positively, "Why not have a license to broadcast? Our country, like Denmark has freedom of religion."

An FM license was granted to Pastor John Nielsen for the city of Nuuk, FM 88.5. Another dear brother from the Faroe Islands, Preben Hansen, an engineer with Radio Lindin, had a very active Christian radio station ministry there. He spoke Greenlandish and offered some fantastic help. This is a difficult language which I am told has a mixture of Inuit, Eskimo and Danish. Galcom set to work to raise funds for the studio equipment as well as 1,000 solar fix-tuned radios. Also, the Faroe Island Church provided a lot of equipment.

With all of the equipment having been sent to Greenland ahead of me, I flew over to Nuuk in September 2008 about mid-month to start working on the installation of the station. Pastor John had opened up a section of the church for the studio. The Communication Department was allowing us to use their tall cell tower for the antenna at a very minimal charge. Praise the Lord! While I was there, we had special phone lines connected up from the church to the tower that would eventually carry the audio signal. Now it was time to begin building the studio. At that time, my twin sister Anne, after a long battle with cancer, was starting to fail rapidly. I had to leave right away if I was going to see her while she was still alive. She was in Chilliwack, B.C. and I was in Greenland about twenty nine travel hours distant by way of Denmark, New York, Toronto, Abbottsford to Chilliwack. When I arrived at Pearson International Airport in Toronto, Florrie met me there with a suitcase full of fresh clothes and I travelled on to Chilliwack. Ann was unconscious by the time I arrived but I spent the last few days with her before the Lord took her home. It is a deep loss when you are a twin but we will meet again.

In my absence, Preben, the engineer from Faroe Islands returned to Greenland with another man to work with Pastor John on completing the studio set up. They did an excellent job of installing all the equipment and the station went on the air March 1st, 2009. What a blessing that was after almost nine years of waiting, praying and working. Greenland now had its very first Christian radio station. Praise the Lord. Since then the people of the city of Nuuk have really appreciated the broadcasts. Lives have been saved and changed and Pastor John's church and the Brethren Church have grown significantly. Praise the Lord!

We even gave out some clear "see-through" radios to the prisoners in Greenland's only prison in Nuuk. Interestingly, the response in the prison has been so great that Pastor John was asked to conduct Bible studies for the inmates. Three prisoners quickly came to faith in Christ and wanted to be baptized. They arranged this to take place on a Sunday. The warden allowed these three men to leave the prison along with eight others who wanted to observe the occasion. The observers all heard the testimonies of the prisoners as they were baptized. It is true that *"... if anyone is in Christ, he is a new creation, the old has gone, the new has come."* (II Corinthians 5:17) What an awesome way God works!

We began working out plans to put some repeater stations in 11 additional communities up and down the west coast of Greenland where most of the population resides.Just one of these stations would be set up on the east coast in Tasilaq. These communities are very isolated especially in the winter: most are over 300 km. apart. There are no connecting roads since the terrain is too rugged. The only connection between them is a boat which travels up and down the coast weekly when the water is open or by plane or helicopter for emergencies. Praise God, we received the licenses in 2011 for four additional Christian radio repeater stations and all at the same frequency, FM 88.5. Nowhere else in the world are we aware of this happening. This made the use of the fix-tuned radios so much easier because they could all be tuned to the same frequency. People travelling from one community to the next could take their radios with them.

After the first station was installed the other communities were interested in getting Christian radio stations into their areas. Now with internet available, we could transmit the programs to the various centres to be rebroadcast locally. Once we were able to have all 12 stations up and running over 90% of the population would be able to hear the Gospel. We started work immediately on the second phase – putting in four repeater stations.

QAQORTOQ, MANITSOQ, SISIMIUT AND AASIAAT

In September 2011, Walt Juchneiwicz, Roelof and son Jeff Datema, Bob Birtwell and I left for Greenland to install stations in these four larger communities. Roelof and Jeff installed the station at the southern tip at Qaqortoq. Bob, joined by Jesper Noer already living in Greenland, did the installation in Manitsoq and Sisimiut and Walt and I worked on the station in Aasiaat.

Raising the tower and antenna in Aasiaat

Each of the stations could reach approximately a 15 to 20 mile radius. It 's just amazing that for approximately $4,500 it's possible to set up a complete radio station. Again, God provided and protected and within ten days all of the radio stations were installed with pole towers and antennas and with their studios connected by internet to Nuuk. This way the radio station broadcasts would be mainly from Nuuk but the setup also allowed for local programming in each community as they desired or if the internet failed.

20

GREENLAND REVISITED

"... And let him who thirsts come. Whoever desires, let him take the water of life freely." Revelation 22:3

I RETURNED HOME WITH THE DESIRE TO PULL TOGETHER some father/son teams to install the remaining seven stations. In August 2012 we had four teams: Robert Kapteyn and son Nathan, Roelof Datema and son Geoff, Robert Booth, whose son was not able to come but paired with Oscar Zollinger and myself with son, Allan, Jr., for the first week and then with my daughter Shari Lou for the second week. Chris Klickerman who was also scheduled to go with us had made up an extra bracket since we were short one but was hospitalized that Sunday and was not able to join the team.

Prior to leaving for Greenland, we met in the Ottawa area for some orientation sessions.

On Friday, Florrie and I worked with them on studio setup, antenna assembly, erecting the tower with the antenna attached, guy wires, grounding and operation of the transmitter. Then on Saturday, Oscar had arranged for the use of a farm to actually set up a working radio station for everyone to get a "hands on" opportunity to see how everything fit together. Florrie also reviewed some cross-cultural information to better prepare them for working with the nationals in Greenland. They felt much better prepared by the end of the day and the sense of anticipation heightened. The next day was Sunday and we had planned to meet for a dedication service at Almonte Baptist Church in Almonte before we got on our way. I understand that this church was founded in 1869 and to Pastor Paul Benson's knowledge they had never sent a missionary to the foreign field. Now they were commissioning a whole team including one of their own members Bob Booth.

The Greenland Team

All of the equipment had been shipped weeks previously and had been directed by John Nielsen, in Nuuk, to the seven sites. Flying to Iceland we were met by Michael Fitzgerald of Lindin Radio and left a transmitter and some radios with him. Then after breakfast we headed to Illulisaat except for Robert and Nathan who were heading to the eastern coast to Tasilaq. They would be installing that station. All the other teams would be installing two stations along the western coast. Transportation is not an easy matter in Greenland.

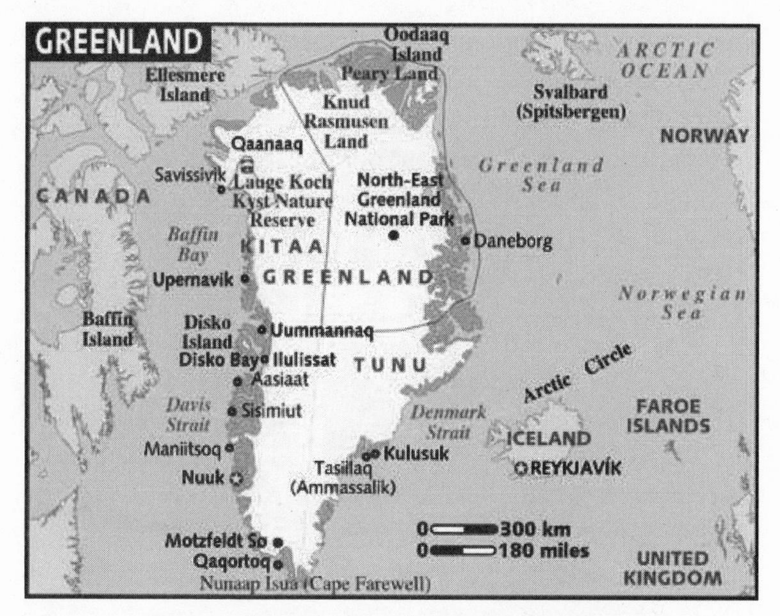

One mother station and 11 repeater stations have now been placed in Greenland

Airplane and helicopter are the most efficient ways to travel but depend heavily on weather conditions. We all split from here with Roelof and Jeff heading south to Nanortaliq, Bob and Oskar heading north to Upernavik and Allan, Jr. and I remaining in Illulisaat to install the station in this community of about 4500 people. Robert and Nathan Kapteyn were the first ones to complete their assignment and they reported on their installation:

TASIILAQ

Josvamik allakkat 1:9
Uangami ivertissimavakkit; tuppallerit pissatsillutillu! Erseqinak nikallornalluunniit; uangami Naalagaq, tassaasunga Guuti, najorpakkit qanoq iliornnitamani.
(Joshua 1:9 in Greenlandic)

"This verse never meant more to me than when my 13-year-old son Nathan and I were in Tasiilaq, Greenland installing a radio repeater station for Galcom International. God proved Himself over and over again throughout our time there. Even when the surroundings and the task seemed overwhelming, God was at work with all of the details.

"We arrived at our destination, Tasiilaq, after two long flights and a short helicopter trip.

Robert Kapteyn and son Nathan helicopter into Tasilaq, Greenland

"We were hungry, exhausted, jetlagged and unsure of how we would accomplish the task of setting up the 52-foot antenna/tower and radio station. Communication was difficult as the people we were staying with spoke very little English. I recall the feeling of heaviness that first night. Nathan and I prayed that the Lord would be in every detail of our work and that night we both recalled God's promise in Joshua 1:9 *"Have I not commanded you? Be strong and courageous. Do not be afraid; do not be discouraged, for the Lord your God will be with you wherever you go."*

"Every night, Nathan and I would pray that verse, and for the Lord to see us through the task we had for the next day. Admittedly, I would still lie awake wondering and worrying about the details and logistics of how we would overcome various obstacles to finish the task. Without exception, during devotions the next night, we could recall how God answered our prayers that very day and brought people and circumstances together to finish that part of the installation.

"Each day's activities brought a new answer to prayer, whether it was through people and expertise to assist us in the installation, or the materials needed at the right time. One particular task of setting up the radio equipment had us installing a grounding connection for the antenna. The quality of the radio signal and safety of the installation is dependent on the electrical integrity of the grounding connection. This was probably one of the most important tasks to get right. When anticipating this all important step, I prayed we would have an approach using material already available such as a metal water pipe or better yet an existing building ground connection. Neither was available, meaning the alternative would be to drive a copper grounding rod into the earth and hope for an adequate ground. This method would be difficult since Greenland is one giant rock. We prayed that we would have success during this step and despite less than adequate conductor material to use for the grounding rod and very rocky land, we were able to drive an iron bar 18 inches into the earth beside the antenna. Our answer to prayer was confirmed when we tested the signal strength of the antenna and it was more than adequate to transmit effectively.

"Nathan and I thanked God that night for the success of the signal tests. We could not find an area in the town where the signal would not reach! We praised God with our hosts as they heard the radio broadcast from the capital city of Nuuk in their own Greenlandic language. That night after Nathan and I gave thanks for answers to prayer, it struck me that people in service are always praying for suitable ground. Not ground in the electrical sense but the fertile ground that Jesus described in the parable of the sower. Matthew 13 *"Still other seed fell on good soil, where it produced a crop—a hundred, sixty or thirty times what was sown."* I pray for the people of Greenland and that the seeds sown there through the Galcom radio ministry may produce a harvest for the Lord.

UPERNAVIK AND KULLORSUAQ

Bob Booth and Oskar Zollinger installed the two radio stations in Upernavik and Kullorsuaq.

Oskar's most lasting impression was that, "When I needed a solution for a problem and I had no option, I was praying to GOD to give me what was needed or a solution for my problem. GOD`s answer was 100 %. I was doing GOD`s work in Greenland. But I do GOD`s work at home too. Why is it so difficult to ask GOD first and not last, when all my own solutions won`t work? From now on, I will include the Lord more in my daily living."

Bob Booth comments on his experience:

"There were a number of very obvious situations where God was at work. We arrived late in the afternoon in Upernavik, Greenland and we had just enough time to survey the property/home where the radio station was to be placed to determine where the radio station and mast/antenna would be located. We had left Canada with just small hand tools although we knew we would need an electric hammer drill. All that was available was a small drill – not nearly capable of doing the job. We had been praying asking the Lord to meet our need and when we arrived at the home where we would be lodging, I asked our host, Carl. if he knew where we could get a hammer drill. Carl answered, "I have a drill I bought 5 years ago in the hall closet." He retrieved the drill and Oskar saw it was exactly what we needed. Carl told us that he had never used the drill and really did not know to this day why he bought it. Now he knew! Thank you, Jesus.

"While in Kullorsuaq I woke up one day feeling sick and unable to eat. Just before noon I told Oskar I needed prayer and I needed to lie down. I woke up three hours later and was totally well. Thank you Lord.

"One night in Kullorsuaq I had a vivid dream. Clear sparkling water was running down the Kullorsuaq mountains and slowly the beautiful water turned muddy. I remember grieving deeply over this change in the water which I sensed involved the Kullorsuaq people. I remember asking the Lord what was going to happen to these people. The answer came as the muddy water started to clear. I felt reassured that the Word of the Lord going out over the new radio station was a key to the clearing waters. What a great harvest there will be in Greenland!

"We were enjoying dinner one night in Kullorsuaq with a Christain family and the young woman who acted as the interpreter told us about a baptismal service one summer evening three years ago down at the water front. As she spoke, I happened to look out the window, and across the large body of water and over the mountain top was an extremely wide rainbow which I pointed out to Dorothy. All of a sudden Dorothy, her husband Moses and Moses' mother all got up from the table got their cameras and took pictures of this rainbow. I wondered why. Dorothy answered this by saying that never in her life time of 22 years did she ever see such a large rainbow. The usual rainbows they see are very, very narrow. The same evening while enjoying coffee and tea, Dorothy shared with us as her husband told of the tremendous hurt he felt 10 years ago when his father left his mother because she had accepted Jesus as her Saviour. I shared a similar feeling from 50 years ago when my mother and father were also divorced. We cried, we prayed and we hugged each other and thanked God for His healing power. The next day, I was told that the mother felt and knew a huge burden had been lifted from her. Thank you, Jesus.

"God was working that weekend as the Lord brought people to the Sunday Service who had not been to church in a very long time. At the Sunday Service and at the Monday night celebrations the presence of the Lord was so strong that people were praying over each other, tears of forgiveness and happiness were streaking down their faces. It seemed to me that healing had started and the waters were already clearing. Thank you, Lord for your faithfulness. It was such a privilege to share with the people in Greenland."

NANORTALIK

Roelof Datema and his son Geoffrey installed the southernmost station in Nanortalik and the most northerly station in Qaanaaq.

"After 30 hours of travel, Geoffrey and I arrive safely in Nanortalik, a town on the southern tip of Greenland, boasting about 1,200 people. We settle into a small one room house and Vilner who lives nearby shows us the faucet in the chicken (his English for kitchen). Water runs not only into the sink but spews down below and soon covers the floor. In his broken English, he tells us he has to go to work and we can fix the plumbing – which we do after mopping up. Later, Vilner takes us out for supper and Geoff especially enjoys the steak and ICE CREAM! Wow, is this Greenland?

"We have prayed for good weather because we are aware of the serious dangers of erecting towers under windy or rainy conditions. Next morning we head back to Vilner's where the radio station will be located. After prayer, I start drilling holes in the 20-foot pipe for routing the antenna cable while Geoff assembles the 10-foot antenna and much of the studio equipment.

"The computer for streaming the radio programs for Nuuk is nowhere in sight; we discover that it is still safely locked up at the post office. The following day, outdoors we attach the antenna to the pole, tie down the antenna wire and from the studio equipment inside check the signal – it's working. Now we raise the 30 foot assembly and attach it to the house. I am sure that the Lord causes me to notice it is not very steady and we use some rope to secure it to the building. Shortly afterwards, a strong wind causes the antenna to lean precariously into the rope. A fallen antenna could be disastrous not only to the antenna itself but for the damage it could cause. We make sure to anchor it securely before leaving.

"Tuesday we set up the computer, hook up to the internet and start the program streaming through the transmitter. Once that is finished, Geoff and I stroll through the town to make sure the signal is strong and clear in every quarter. What a terrific feeling to hear the broadcast everywhere we go even up into the foothills of the mountains. It's amazing how far the signal travels. Praise the Lord. We give Vilner instructions on how to maintain the station and our work here is completed.

QAANAAQ

"Tomorrow we head out early to what is known to be the most northerly church in the world, in Qaanaaq. Foggy weather delays our series of flights many hours and we are wondering how we will get our once-a-week connecting flight. After many hours of delays and two missed connections, God gets us to the Land of the Midnight Sun and after staying up until midnight to verify the fact, we head to bed thinking this install will be easy from a house high on a hill. Surprise! In the morning, we discover that Tukumeq, our hostess, and her husband, Arqinuaq, do not own this house but rather have a small building at the bottom of the hill – not the best location for an FM transmission. On further investigation we find that in exchange for an elevated transmission site, this building comes with a big pile of stuff, mostly outside: wood, tools, metal and junk of all descriptions. Now we need to be able to raise the antenna high enough to get a clear, strong signal."

Geoff takes it from here: "Last night, Dad and I went out for a walk at about 10 p.m. in broad daylight and met Leo, a nice Greenlandic man who speaks very good English. Apparently, he used to follow the Lord but developed a drinking problem and no longer cares for Christian things. We told him about the radio station and he says he will look forward to it working. I think, maybe the Gospel will reach him and he will come back to the Lord.

"Anyway, the next day, while I start assembling the antenna and putting the studio equipment into the little radio house, Dad is figuring out how to mount the antenna. He finds locations for the support cables for the tower but needs some metal rods and a big hammer. He finds Leo who lends him a hammer and points him in the direction of a friend who would be able to supply the metal rods. The friend ends up giving us the rods for free! So thank God for Leo and his Swedish friend for supplying what we need. We also need some pipe long enough to hold the antenna up as high as possible. Dad leaves the house with no idea where to find a length of pipe. He returns to the place where he obtained the rods but no one is there. He starts praying and a man shows up in a truck. His English is very poor so he phones Fritz who gives Dad a ride in his truck to go see Per who has pipe available. For 500 Krones he purchases the pipe he needs. Another answer to prayer.

"All the time we are working we are waiting for the telecom people to install the internet into the little house. They keep saying, "Tomorrow". If it's not installed in time, we'll have to leave the site before completing the hook up. Well, another answer to prayer: the man shows up just in time for us to stream the programming from Nuuk.

"In the pile of "stuff", Dad is able to find two bolts long enough to fasten the pipe to the wall, but they need nuts. He asks me to see if we can find nuts to fit them – our Greenland version of finding a needle in a haystack. "God, I need to find a nut of exactly the right size and thread in this junk pile." Within 30 seconds, I find the first nut! This is amazing and encouraging! Now for the second one. After about 10 minutes of searching, I am getting a little frustrated. "God, I just can't find the second nut... help." Then I see this yellow plastic bag with some nails, screws and a dirty old sock in it. I think, "Ha ha, no way there is a nut in there... but I'll dump it out and look anyway, just for fun." So I dump out the bag and there it is, the second nut... exactly the right size! Again, God answers prayer.

"The next day, we set up the antenna temporarily to test the signal. Even though it's not high enough, the signal reaches quite far. We're excited. Arqinuaq has just returned from hunting. He and his brother caught three narwhals. I can hardly wait for supper!

"With the help of some men from the small church here and more parts from the junk pile, we get the antenna up in the sky. Thankfully the weather has been fantastic – no rain, wind or snow so it goes up safely and quickly. Dad starts streaming the program by internet from Nuuk and makes a few adjustments to get the optimum sound ready for church tomorrow. In the little time we have left, we will be training people on how to use the radio station and how to prepare their own programming. The Christians here are so happy and thankful for the new radio station. It's amazing how joyful and content these people are because of Jesus, even living in this dark part of the world. Please pray that the Lord will begin preparing the hearts of those who desperately need him, that they will listen to the radio and by faith in our Saviour be changed forever.

"During one of our last days, Arqinuaq takes us in his boat to see some narwhals. We saw lots! He tried to catch one a few times with no success even though they are usually easy to catch. All you need to do is boat out into the fjord and watch for a narwhal to surface to breathe. When you see its black bump/hump allow for five to eight breaths, hop in your kayak and paddle to where you think he will appear next, harpoon it and pull it back into your boat. Timing is everything. Needless to say the advantage goes to the narwhal who can also spot you on the surface and avoid coming up in your proximity. Every once in a while a seal would pop up beside the boat and look at us. We could hear it breathing, it was so close. Cute little thing, just bobbing up here and there looking around like a prairie dog then disappearing below the water just as quickly. We also went by boat with Tukumeq to the most northerly town in the world (apart from scientific communities) with about 40 houses. Expecting to find at least six feet of snow, we were surprised to see it was just like Nanortalik on Greenland's southern tip.

"We finished up our work by training our host and hostess on how to use the radio equipment, packing up and saying our good-byes. This has been quite an adventure and we praised God that both of the stations were working well as we headed for the long trip back home."

Roelof later wrote these impressions of their trips:

"What a blessing it was to work with the Galcom team on the radio installation project in Greenland. When we spoke with Allan at the Missions Conference at The Metropolitan Bible Church in Ottawa in November 2010, we had no idea where The Lord would be taking us. I never thought I could be a missionary because I'm an engineer. I could support missions, but how could I use my technical skills? It is now clear that through my education, career and travel experience, God was preparing me for a unique assignment with Galcom where I was able to use my technical skills for kingdom building work.

"Traveling and working with my son who was 16 at the time of our first trip, was a true blessing for our family. We went to Greenland twice for a total of five weeks to install radio stations to broadcast the Gospel message in the native language of the people in remote communities in Greenland. The work was rewarding because the people were so grateful. They anticipated our arrival and celebrated the completion of each installation. They told us that because of the Gospel being sent out over the radio, people would be alive in the spring who otherwise would have given up hope during the dark winter months of the far North. We had the privilege of installing a radio station in the most northerly town in the world, Qaanaaq.

"One of the blessings of missions work is the opportunity to see and experience other parts of the world and to meet brothers and sisters in Christ who live in circumstances so different from ours. Having been on two trips, this has become a part of our life. We welcome the opportunity to do it again."

ILLULISSAT

Allan Jr. and I were met by Pastor Joseph, took a few minutes to get settled into our accommodation, and then set out to evaluate the location and equipment. Tuesday morning we started in earnest and discovered very quickly that the brackets for securing the pole to the house were much too small. This became a matter of prayer: where would we find brackets that would be suitable? In the meantime, we assemble the antenna and attach it to a shorter piece of pipe to test before raising and anchoring it in place. To our dismay, the custom pretuning of the antenna is faulty. We immediately think of the other three teams and whether or not they are facing the same problem. We set up the studio and transmitter and find the antenna is tuned to 91.1 instead of 88.5. Allan, Jr. is quickly figuring out the calculations for changing the three interconnected variables on the antenna so he can relay the information to the other teams. Four hours later, he has the antenna working. What a blessing to find that this is the only antenna with this particular problem and that it just happened to be at our location where only Allan, Jr. knew how to correct it!

We had managed to get some brackets made up and proceed to erect the tower. About half a dozen men show up at just the right time to lift the tower into place. All the connectors for the RF cable are well sealed and the guy wires are in place ready to be tightened. Praise God the raising of the antenna goes well. Having fifty feet of metal high up in the air always poses its dangers. We anchor the tower to the house and get the three guy wires tightened to secure the whole thing snuggly. We head into the studio to turn on the transmitter and watch as it powers up to 45 Watts. A few minor adjustments and the station is up and running! Only a few people in this community know Christ as their Saviour: now they will all be able to hear the Gospel message on a regular basis.

It was such a privilege to work so closely with my son on this project especially since he designed and built the cornerstone transmitters we were using at each location and did all the testing to have them FCC approved. After all of that background work, he was excited to be able to install one of these stations himself. Pastor Joseph, in the Sunday morning church service, took time to dedicate the new radio station to the Lord. Allan and I packed up and headed to the airport where we met Shari Lou who had just arrived via Iceland. Allan had to get back to Hamilton but Shari Lou would accompany me to Uummannaq to do our second station. Allan Jr. writes his impression of the trip:

"My back hurts from two days of travel. The aircraft manufacturer couldn't possibly have engineered the seatback to be any harder. The relentless buzz of the twin turbojet engines irritates my knee which is wedged between the aircraft hull and the seat ahead. Then I look out the window and see mountain after mountain protruding up from the ocean depths encircled by glassy blue water and gleaming white icebergs. In the distance the rugged coast of Greenland leads into a seemingly endless mass of luminous white glaciers. My petty complaints and grumblings put me to shame. They quickly melt away and my first glimpse of Greenland leaves me awe struck by the beauty and ruggedness of God's creation.

"We arrive at Illulissat airport and my father and I see a Greenlandic man with an immense smile. My dad says "Hello, are you here to pick up Pastor Allan?" His smile doesn't fade but he says nothing. My dad repeats the question and in broken English the man replies "Pastor Allan, yes, we go". We climb up into his old van and he proudly drives us through the winding streets of his small town. I look to the left and see towering stone cliffs. I look to the right and see small boats navigating around icebergs the size of skyscrapers. The small houses are painted bright colours. There's an abundance of sled dogs which surprisingly outnumber the town's 5000 residents. The van ride ends at a two-storey red church neatly perched on the side of a rocky slope. I can tell that my dad is focused on the radio station because he immediately looks through the building searching for a suitable location for the studio and antenna. Our Greenlandic host looks puzzled but his unending smile continues.

"I was warned that north of the Arctic Circle, August has some cold days. This was one of them. The fog and rain hide our view of the ocean. We climb up the steep rock slope behind the church looking for an appropriate tower and antenna installation location. The rain is soaking everything and making the rock precariously slippery. I step on a small patch of moss and it rips off the face of the rock sending me sliding down about four feet. Luckily, a large rock breaks my fall. Fear is burdening us because erecting a forty-foot tower will be an extremely dangerous job. We take time to stop and pray. We're praying that the weather will clear up for the duration of the install.

"The morning sun greets us as we layer on our clothing and venture out into the cold August arctic air. The sun is banishing the rain, and the rock is already drying up in some places. It is turning into a beautiful day. We rejoice in our first answered prayer. My dad and I both agree that the best antenna location is behind the church. I dart between the possible guy wire locations measuring their distance from the antenna tower. The warmth of the sun is welcoming. I remove layer after layer of clothing. Could the weather be any nicer? It's certainly not what we expected.

"I am very proud of how hard my dad is working. He has a focus and determination that is exceedingly rare. He has an energy level that a twenty-year-old would be jealous of. He also has a natural ability to motivate people, making them quickly aware of needs and eager to lend a hand. Our smiling host arrives. My dad says "We need some help here. Do you have anyone who can help us tomorrow?" Our host gives us a big smile followed by silence. "Can you help us tomorrow?" There is silence. This is not going well. My patience is waning and the clock continues to tick the minutes away. Finally, our host gives us a nod and says "Yes, people tomorrow". I hope he really understands. At least he's still smiling.

"It's time to raise the antenna tower. This is the most dangerous part of the radio station install. There are countless opportunities for catastrophic failure. Our host arrives just in time with a group of men from the church. The language barrier makes communication difficult and could possibly be the most dangerous factor. The ten-foot antenna atop the forty-foot tower lies on the ground looking strangely out of place amidst the rock. The guy wires are carefully attached to the mast and rolled as neatly as the terrain will allow. I look around at the motley crew of eager radio station installation engineers. My dad is shouting instructions at some of them. We try to verify every instruction making sure that they really understand what is required of them.

"The men start pulling two of the three guy wires. One group lifts the antenna. Little by little it rises into the clear blue sky. Two of the men clamber onto the church roof steadying the mast. If the antenna moves past top dead centre and the third guy wire is not adjusted correctly, then it could topple over and injure someone or break the precious antenna. While tightening a clamp I see that one of our first time engineers is loosening the guy wire instead of tightening it. I yell at him to stop, but he does not understand. Two guy wires support the antenna but the third is now wafting in the breeze. If the wind shifts direction then the tower could fall. I try to have one of the men who speaks a little English translate the instructions, but it's hopeless. I am sixty feet away. My heart is in my throat. As quickly as I possibly can, I finish tightening my guy wire and rocket up the steep hill to help secure the wire. It's tightened just in time. I take a deep breath and relax now that the tower is secure. The tower and antenna are raised and they look beautiful. We all share a wonderful feeling of accomplishment.

"It is 10:30 at night. I am tired from the day's work. I am sore from the countless trips up and down the side of the hill assembling the radio station. The sun is setting and has been setting for about four hours now. This may be my only chance to see a Greenlandic sunset. I leave the church walking down the steep streets. A sled dog is sleeping on top of a rustic doghouse. Arriving at the edge of the ocean, an eerie blue and orange sky reveals a bay that is a visual symphony of colour and shape. Icebergs and their mirror image dot the glassy tranquil water. The endless horizon pulls my eyes to the distant silhouetted coastline. Wisps of translucent cloud dress the sky. I fill my lungs with the most crisp, fresh air that I have ever experienced. Filled with an immediate sense of peace and calm, I wonder if this is perhaps a faint glimpse into God's infinite splendor and majesty.

"Encouraged by the previous day's successes we start work on the studio. My father is connecting the studio equipment while I connect the computer feed from Nuuk, the capital of Greenland. Praise God, the radio station works! Three ladies from the church arrive at the studio. One of them speaks English very well and says "We heard the radio station transmitting so we came down to the church to see it". She starts to cry. I don't understand why she is crying. She says "This means so much to us. Now many people in our town will be able to hear about Jesus." The tears are dripping onto her blouse. I realize how much this little radio station really does mean to them and how much they care about their community. The Greenlandic people are extraordinarily giving, exceptionally kind, and incredibly friendly. They face many struggles though. Suicide, physical and sexual abuse are an ever-present reality in their small towns. The ladies, my father and I start praying that this station will be a blessing to their community and a light to many people.

"I make one last trip up the harsh rock hillside. At the summit of the rock someone has painted a white and black circle. Standing on the circle, looking to my left, I see the little red church with its shiny new radio station antenna raised slightly higher than the peak of the hill. In front of me the ocean floats its massive icebergs. To my right an airplane lifts off from the short runway. Soon I will be saying goodbye to my smiling host and leaving on one of those airplanes. I'm encircled by the quiet town and I have a chance to reflect on my short time here. I am very grateful that my father and I shared this experience. The GPS coordinates of this extraordinary location are 69.217531,-51.110303. However remote the chance, I hope that someday I'll be able to return to this precise spot with my family. Just as the Greenlandic people in the town of Ilulissat have left a permanent mark on my heart, I pray that this radio station will have a lasting effect on the hearts of these wonderful people."

Allan Jr., left for Nuuk to head home and Shari Lou arrived in Illulissat. On Monday we are to fly out to Uummannaq, but when we arrive at the airport our flight has been cancelled due to fog. With only about five days left in our schedule a delay would be disastrous since planes do not fly every day. We're told our plane might leave later in the day. We leave the airport and head back over to the church where my son and I had installed the radio station and pray. Then we head back to the airport, and just as we are going down the stairs (there are stairs going everywhere in Greenland!) we see over the stairway a beautiful rainbow shaped arch of fog – yes, fog. We take it to be God's promise that He will get us to Uummannaq today. We stop at the hotel to pick up our baggage and are summoned to make our way immediately to the airport. There we find seven other people also waiting for the same flight. Amazingly, Air Greenland has ordered a helicopter that holds nine passengers and they fly us the entire distance to our destination. What beautiful scenery we are privileged to see from that vantage point.

Two women run the ministry in Uummannaq both named Ane Marie. Ane Marie 1 met us at the airport and had supper ready for our arrival. The studio was to be housed in a bedroom of their home. We went to check out the equipment. To our surprise, Ann Marie's teenage son had already started to set up the equipment as best he could: he was wise enough not to turn anything on. The arrangements were that each location would have certain power tools, brackets and a 40-foot steel pole to support the antenna. But, let's back up and hear Shari Lou's account of her experience.

UUMMANNAQ

"God in Greenland: It is 11:40 at night in Illulisaat. I am standing with another young woman outside an old Danish hospital on the west coast of Greenland watching the sunset over the ocean where it will briefly hide from the land and quickly make it's reappearance the following morning. There are icebergs peacefully floating by, too many to count. Ships can be seen navigating their way past them towards the shore, casting an eerie light on the icebergs as they pass. The sky is a wash of colour – oranges, light blues, and a deep dark indigo high up into the sky. There is a freshness in the air which I had never felt before. Like the crispest spring – the kind of air that make newly born colts jump into the air in joy.

"Greenland has a way of seeping into your bones. I believe we not only know God through the writings telling us about His Son and His people, but we also know him through his creation – the plants, animals and especially the people. In different places we get to know God from a different perspective. In Greenland, God seems to be communicating through the extremes found in that place. There's a harshness in the vast centre of the country – packed-snow covered rock, while on the edges there are Greenlandic blueberries by the billion creating a low blue-green cover as far as the eye can see. I could see Him on a mother's face as she dressed her child for the outdoors. I also saw Him in Ummannaq when I watched a broad smile abruptly break into sorrow while reminiscing about a fond memory of a brother who had later committed suicide. Greenland taught me about this side of God. In a place where your body and soul can be eaten up with the ever consuming needs and the harsh climate has no mercy, it is not possible to navigate this life without being within the community.

"The station my Dad and I are to assemble in Greenland is on the small island of Uummannaq. As we come closer in the helicopter, I am stunned with the beauty of it – like a precious stone on the beach. The central mountain rises up from the deep blue of the ocean – grey strips through coral-oranges, with the pink-peach reflection from the sun making it appear like a source of light itself. Around the mountain there is a disc-like flatter area where cheery houses stand of all different colours: canary yellow, electric blue, bold red and grass green houses glittering along the shore. In Uummannaq there is no dirt, leaving wires and pipes exposed and the little dirt that is there has been brought in from ships to allow the dead to be buried according to tradition. There is nowhere to go when there is no ice to freeze a pathway to the mainland, unless one takes to the risky business of navigating icebergs in the waters.

"The previous 3 days have been spent stuck in Illulisaat while flight after flight has been cancelled due to fog and weather conditions. It leaves me wondering about the place we were attempting to get to –Uummannaq. It sounds so similar to the other Greenlandic words I have learned. This morning, Air Greenland made special arrangements to have us helicoptered up the coast to the island – usually this trip consists of a short flight followed by a several minute helicopter ride - and informed us about 20 minutes before the flight that if we wanted to go we had better be at the airport when it took off. We leave the ground in the red shiny helicopter, hovering over land that is rarely seen by the human eye. The pilot is Greenlandic, a people group similar to the Inuit in Canada. He looks very accomplished, wearing his leather, sheepskin-lined jacket : proud to be flying for his country. We travel over deep antique-glass green lakes, watching icebergs the size of city blocks float silently down the coast. Landscapes that look like they should be on another planet pass by us – I look at my father, who has seen so many countries and so many sites, and there he is, his jaw dropped with the awe of it all.

"As we land, and thank the pilot, it hits me that I am already in love with Uummannaq. Greeting us with smiles are two Ane Maries who administer large warm motherly hugs and smiles. The Ane Maries are also Greenlandic and very excited to have my father and me stay with them in their bright blue house atop a hill.

"Setting up a radio station made me feel like a kid with a new Lego set: so many pieces, and they all need to work together. In the living room I brought out the box with the antenna in it. In the house that evening was Ane Marie 1 and 2, Ane Marie 2's son, and two young girls, one named Arnartaq and her little sister Nielsine, probably 10 and 12 who were staying with Ane Marie 2 because their parents were "preoccupied" with having a good time. I spoke perhaps 40 words of Greenlandic, and they had twice as many words in English, so communication was rather basic, but Arnartaq and Nielsine sat enraptured watching me start to assemble the antenna on the living room floor of this brightly blue coloured house at the top of the hill before the mountain seriously started. I pointed to Arnartaq, showed her how I tighten screws and soon she was tightening them on her own. She did it with a focused, aggressive movement, and then turned to me for approval of her job with a smile. I nodded and I had little Nielsine do the next screw. Next thing they were nearly fighting over each piece we had to assemble to do the work. We were just about done when it was time for dinner. We were seated at a relatively long table with an even longer sunset happening outside. It was summer, and the crisp air came in through the crack the window was open. For dinner it was a prepared frozen peas and corn with rice and fish fingers. I had prepared myself for the whale blubber soup and seal meat I had heard stories about from others who had done a similar trip.

"The days passed, the radio station slowly went up. I prayed that the work we were doing would impact the lives of the people here. I prayed that Arnartaq and Nielsine would be protected and loved. I prayed for the people who had lost family members due to abuse and suicide. Sexual abuse was a real problem in these small communities, with victims living houses away from their abusers for their entire lives. This is a place where problems don't go away, they need to be lived with. The harshness of this reality brought me back to appreciate God's grace and renewal; the only way forward for those broken lives – and for my life as well. The radio station we were putting up would be truly a voice of hope, and a living option for how to navigate through the storms.

"The way these operations ideally work is that all of the needed local supplies are assembled for the teams before they arrived. We ran into our first hitch when we realized the base pole would not be appropriate for the forces needing to act on it. We also realized that we didn't have the sturdy clamps we needed to attach the pole to the house so it could act as a main support. These were big issues that could halt our work and make our trip stunted at best.

"My father and I put on our jackets, opened the door and stepped out into the outside world. Mickey and Princess, two sled dog puppies were overjoyed to see us, partially because I had decided they were the most adorable puppies in the world, and lathered attention on them any time I got a few minutes. My brother had warned me not to pet the sled dog puppies no matter how cute they looked – advice that lasted about 30 seconds once I saw these two – I suppose I am in veterinary medicine for a reason. The feeling as we left the house was that we were both open to adventure. We had a need, as we left the house we asked God to meet it and set out on our search. Neither my dad nor I seemed to have much trepidation going into a foreign situation like that.

"The view once again hit me as we walked down steep streets lined with tiny houses, down and up rickety staircases. We came to the only mechanical shop in town. They said they might have a few pipes out the side of the shop. We found one that was promising. There was mention of a man named "Kim" that was working down in their other warehouse. We walked for about 5 minutes around a large rocky edge, and saw a junk pile, with pipes across the street. We looked through them carefully, and our answer came: we had pipes that would work for our antenna, we would take two and thread them together to create a 40 foot length that would tower into the sky. Across the street in the junkyard, we took a cursory look for other parts as the two sled dogs nearby watched us warily from where they were chained to the rock.

"Walking back towards the second mechanic shop, I saw how thrilled my dad was that we found this pipe, and what an answer to prayer it was. At the shop a sturdy, strong Danish man, we thought was named Kim, looked in his element. He greeted us in the little English he knew and I could feel him watching me intently. As he guided us around the shop looking for the pieces we needed, it became clear that we would need something custom made. He told us he would fashion something for the bracket attachments and another problem was solved. On the way back up the hill carrying the pipes, it seemed like another hurdle was overcome and this station might really go up.

"On our last night in Uummannaq, Ane Marie 1 was busy in the kitchen. Walking in I said to her "One thing I always wanted to do was to make my own mittens. Do you think you could show me how to make them?" Our work putting up the tower and setting up the radio station was done for the night, and with the extra time I thought I would try to learn something new. I was ready to buy the fur; I just needed a little direction in making them. She looked suddenly excited and energized, and the tea she was making for herself was quickly forgotten and left steaming on the counter. She had a big cupboard open that I had never noticed before and was taking boxes out with sealskins, mitten patterns, leather needles – items from a different place and time. We carefully traced out the pattern and started first with our red felt lining which was carefully cut, then we repeated with the sealskin, but the opposite way, so that the seams would fit together.

"Surgically, I had spent time stitching together cows in various situations and knew how tough their hides could be, but watching this woman throw a needle through dried sealskin leather like it was silk was impressive. I had her on my list as a good candidate for being a large animal surgeon if she ever wanted a career change. As we worked together, sitting there side by side in her tiny kitchen, talking and laughing with a hot cup of tea beside us, we came to talking about her community and the people.

"It was one of the times I would get a better impression of Ane Marie and her struggle. I had become really fond of her over the days we had spent together, and images of her bent over a bucket of fish with the sled dogs waiting hungrily were held dearly in my mind. She had a strength of conviction and of lifestyle that few other women had. She worked very hard, and made her home a welcoming place. As she began to speak, I listened intently to try to understand her. "I had a dream that Jesus appeared to me, like a bird. I felt that he had full and complete love for me and I could look at everything differently from now on. Sometimes as I am going to work I see birds lining the stairs and I am reminded again of the blessings Jesus has given me. And the Lord says to me, it has been 2 years since I saved you" She had this unabashed carefree way of talking about what He had done for her life, and what it had been like before she had become a Christian. I could tell that this radio station meant the world to her, because she knew Christ could make a world of difference for the people she cared for in the community. It was so incredibly peaceful talking to her there while we worked.

"Thinking back to the events of the day as we put up the tower, brought back the previous feelings of concern for the team. That morning, in the clear light of day on a rocky surface just on the far side of her house lay a 40 foot tower along the ground, the erection of which I had been truly fearful. I did have enough good sense to know that without enough people and physical support, it would be easy to lift up a tower like this past its point of equilibrium and have it swing back the other way, hitting someone in the head, damaging a house – any number of outcomes. Without the right support at its base and clear access to the guy wires it was a dangerous situation. Even with those things it was a dangerous situation – and the only people we were assured to be there were myself, my father, Ane Marie I, Ane Marie 2 if she could leave work, and possibly Arnartaq and her little sister Nielsine.

"So the previous day we started on a campaign to enlist help from the friends we had made over our brief time in Uummannaq. I had started with Kim and he ended up bringing a friend of his as well. An ambulance driver at the hospital where Ane Marie worked agreed to come. Ane Marie 2 managed to make it there, although her son had left for Denmark where he was attending school. Arnartaq and Nielsine were out there helping with fearless determination.

"One guy wire was angling down the hill in the open with a steep drop just past where the hook was to secure it. The other hook was very close to their neighbour's building, and the final one was out in front of their house near the road. The tower would stand bolted to the back corner of Ane Marie's house. The plan was to angle the tower so the base of the pipe was right at the corner of the house and to carefully lift towards the house with the guy wires. As it got higher up, someone would need to make sure the base was securely in place to avoid it swinging out. We started and my prayers suddenly took on a new urgency. "God please protect the people working on this – I know you can work in this community and their lives through this station. There are dark forces at work here that only you can overcome.

Help us get this tower up and the station running." I found my prayers became simpler and more to the point the more I could feel my heart in my throat. We assigned people positions and started pulling with the guy wires, my father and I moving from one position to the next to make sure that nothing was going wrong. As we got up about half way, one of the men got up on the roof of the house and with an extra wire started pulling from there. It started rising up, 10 feet off the ground, 20 feet, and at 30 feet its critical point my main concern was maintaining control. Reflecting on this a year later, I notice that my hands are sweaty and tense just with the memory.

"As I looked up from my point at the corner of the house, I realized the 40 foot pole was now straight up and down. We needed to tighten the brackets to the house and fast. Kim and I started tightening bolts – I started from the top and he from the bottom. Without his strength it would have been a huge challenge. He and his coworker seemed to have had experience with projects of this magnitude from the confidence and intelligence they used to work in the physical world. I used leverage at certain points in order to line up the holes for the bolts, knowing I'd have to make use more of physics than my physical strength.

"As we were tightening the last bolt to clamp the tower, the guy wires were being tightened, at first temporarily and then in a few minutes they would be attached permanently. I could let go of my post and help at their connection points. I stood with my dad at the last guy wire to be tightened and he looked up at the tower just in front of it. As he stepped back he tripped over the wire, and I caught him just before he fell and pulled down the wire with it. If for any moment of the trip I was supposed to be there, I felt even if that was the only useful thing I had done, it was enough. "Thank you Lord" I prayed as I exhaled.

"We tightened the last guy wire and thanked the volunteers heartily. The feeling of accomplishment was almost palpable as we looked up at this perpendicular tower and antenna. A quick test in the studio inside confirmed things were working and once the license came through we could start broadcasting. That evening as we prayed over our giant red fish that was to be our dinner (Ane Marie seemed excited at my interest in Greenlandic cooking while my dad perhaps was less impressed with our foray into more "authentic" cuisine), we thanked God for His presence in this community and His love for us, His children. Fear had no place in the room that night. Something special had happened there, and I had been part of something good.

"We all have our perspectives when it comes to mission work. I was raised as the child of a pastor and of missionaries, so I knew better than most the ins and outs of what happens when one is sharing their faith. I have always believed in my father's mission; we need to share Jesus first and only, learning and understanding the culture and context his teachings would properly fit into. Knowing that it was Greenlandic people who invited us there, who would be running the station, and who would be benefiting from the message certainly impressed me. Since we left Greenland, the people and the station there has often been on my mind, and there have been times when I am outside looking at the stars, wishing I could be on that coast in Uummannaq watching the icebergs silently float by and not seeing fear, seeing God."

As I (Allan) head back home, I am overwhelmed with God's goodness. The long years of praying and planning for Greenland have begun to bear fruit. The little solar-powered, fix-tuned radios are distributed and the precious message of Jesus' love is penetrating the frigid airwaves of one more frontier.

In my mind's eye, I see people of every nation, language and tribe streaming from every direction exuberantly towards heaven. They are singing and rejoicing as they praise their Saviour, the One who gave His life for their redemption. Among their number is José from El Salvador, the obnoxious drunkard who first heard the Gospel through a GALCOM solar-powered fix-tuned radio. The Colombian General is filing by with many of his soldiers following: they too heard the Gospel through the special green dual-band radios. From another direction comes a group of blind Filipinos still grasping their radios as they sing. Believers from Sierra Leone, a huge band, all excitedly praising God are dancing in rhythm. From Moldova, children by the hundreds are joyously running and clapping their hands as their care-givers accompany them. African, European, Asian, South American, North American, Australian believers: the parade stretches as far as the eye can see.

And I am thinking of the thousands of godly men and women around the world with whom I have been privileged to work. Together, we have sought to bring the message of the cure for sin through the death and resurrection of our Lord and Saviour, Jesus Christ. And we have seen and experienced the power of this Christ to change lives: every country, language, tribe, nation – God's Message is powerful to all who believe. Praise His Name!!!

21

EYES ON THE FUTURE

"And the things that you have heard from me among many witnesses, commit these to faithful men who will be able to teach others also." II Timothy 2:2

STARTING BACK IN THE LATE 1990'S, I BEGAN MENTIONING TO the Board of Directors the need for someone to work into the role of leadership for Galcom International, Canada. Although I was blessed with good health and tremendous energy, I did not want to presume that to be the case indefinitely. We prayed earnestly, considered a number of applications and interviewed several people. No one seemed to come on the scene who had a passion for the Lord and this ministry and the call of God to carry it forward. Years went by and we continued to put out feelers.

In 2007, I was preaching at Faith Gospel Church in Hamilton, one of our supporting churches since we first entered missionary work. At the close of the service, I actually was leaving the pulpit when the Lord prompted me to mention that we needed prayer for a new, younger director to prepare to step into the leadership role of GALCOM. In the congregation that morning, as usual, sat Tim Whitehead. His thoughts were suddenly jolted.

He distinctly felt the Lord's call on him to pursue that role. He immediately whispered to his wife, Melody, "We have to talk after the service."

Tim had been serving with 100 Huntley Street in the finance department and then as the National Administrator of the Circle Square Ranch Camp ministry. As he met with me and then with the Board of Directors, it was evident that God had chosen this young man to fill the position. We had watched Tim grow up in the church and excel through the Awana program, eventually taking serious leadership within the church. God had been preparing Tim long before we started looking.

Tim spent two years with us initially learning the ropes and shadowing me day by day. The second of these two years he began to take on the leadership role as Florrie and I stepped back from administration and moved largely into the area of being Galcom representatives. By 2009, Tim was ready to shoulder the task as the new Executive Director. God has blessed him and continues to lead him as he takes the helm.

Ken and Margie Crowell, in Tiberias, Israel, were heading up the MegaVoice ministry making audio Bibles. Charles and Judy Cibene, their son-in-law and daughter were working with them. Ken's mantle was passed to Charles to head up the MegaVoice ministry as Executive Director. Ken, sadly for us and his dear wife, was taken home on January 25th, 2012.

At the same time, Gary Nelson, president of the US office of Galcom, was also looking for someone to fill his shoes. Administration of the funds donated by Harold and Jo Ann Kent was a large part of this ministry. Tom and Stacy Blackstone had just returned from missionary service in Turkey and soon heard about the Galcom ministry. Before long, Tom was working into the leadership of Galcom USA as their new Executive Director. Just recently, on January 23, 2014, Harold, one of our founders, was welcomed into his eternal home.

God has graciously provided three godly men, along with their wives, to take up the reins of Galcom International and MegaVoice.

The new team: Tom & Stacy Blackstone, Charles & Judy Cibene and Tim & Melody Whitehead

We marvel at His leading and rejoice in His goodness. God raised up this ministry in the beginning, bringing together three men from three different countries who had never known each other previously. That three-fold cord has endured and is now intertwined with a new three-fold cord. As God's word says, "*A three-fold cord is not quickly broken*" *Eccl. 4:12.*

By prayer, together, we will continue to pursue the calling God has placed on us to send out the Gospel message on **waves of hope**, "***Looking unto Jesus,*** *the author and finisher of our faith, who for the joy that was set before Him endured the cross, despising the shame and has sat down at the right hand of the throne of God." Hebrews 12:2*

ENDORSEMENTS

The Galcom radio is a wonderful tool that has caused GOD'S KINGDOM to grow all around the world. Time after time, after we would put in an FM station, Allan would come along and provide the radios to the listeners, and the Kingdom would grow. The faithful, generous, gracious service of Allan McGuirl has been an inspiration to many as well. He has been an excellent partner and representative of the King. In this book you will read how God leads and blesses his people and how He uses them to build His Kingdom. My prayer is that this book will challenge you to allow Him to do the same in your life.

Dr. Ron Cline, HCJB Global

The story of Galcom is one of faithfulness and faith. Allan McGuirl is a consummate story teller and the subject of these stories is how the Gospel breaks through tough barriers in innovative ways. I highly endorse this captivating and compelling book.

Dr. Bill Fietje, President of the Associated
Gospel Churches of Canada

This is the remarkable story of Galcom International, and of Allan and Florrie McGuirl, its founders. Behind every work of God is a human story of people who are, "fully persuaded that God had power to do what he had promised" (Rom 4:21). From very small beginnings Galcom has established gospel radio stations, providing fix-tuned radios in abundance that are permanently tuned to the gospel stations on most continents. I have seen the enormous effect of this in different countries, and our congregation at The Peoples Church in Toronto has been thrilled many times with the impact Galcom is having. Here is their story that is enjoyable, enriching and inspiring.

Dr. Charles Price, Lead Pastor, Peoples Church, Toronto

What a captivating and faith-stretching story. My heart was inspired by the passion and perseverance of Allan, Florrie and the entire Galcom team in broadcasting the good news of Christ Jesus to those desperately needing to hear it. Jesus promised to be with those who seek to make disciples of all people groups (Matthew 28:18-20). Galcom is living proof that Jesus is still keeping His promise.

Dr. Rick Reed, President, Heritage College and Seminary

This is a story that was waiting to be told! The author is as unique and influential in the Kingdom as the mission God gave him. Allan has been showing and telling his ingenious devices on national television with unabashed enthusiasm and child-like delight-- complete with joyous hoots and 'woo-hoo's!

Galcom's fascinating history chronicles a God-given vision to boldly go where man cannot go with the life transforming message of Jesus Christ. Relentless ingenuity and brilliant strategy have been driven by a love of the Lord and people. Any conversation with Allan will conclude with a heartfelt prayer modeling what it means to simply trust in Jesus and enjoy life with Him! The faith lift is inescapable. Here is one of the great missions initiatives of our time. Thanks for your obedience and faithfulness Allan!

Moira Brown, Host on 100 Huntley St.

Waves of Hope is a powerful reminder to followers of Christ and lovers of the gospel that nothing is impossible with God. It's the amazing story of what God can do through a man who is totally dedicated to sharing Christ with the world. But this is more than an inspiring biography, it's the story of Galcom International, a ministry with humble beginnings that today literally reaches thousands of people with the gospel all over the world. If you are interested in the unique and incredible ways and means that God uses to spread the gospel this is a book that will thrill your soul. As Allan McGuirl's pastor I have been richly blessed by the many times he has placed his hand on my shoulder and prayed for me. Reading his book has powerfully reinforced my belief that the gospel is the power of God for the salvation of everyone who believes.

John Mahaffey,
Lead Pastor, West Highland Baptist Church, Hamilton, ON
Council Member of the Gospel Coalition

"One of the clear principles of Scripture is that, for the Christian, all of life is preparation for what God will have us do next. The life of Allan McGuirl exemplifies this more than many. God called a talented and determined young man to a great kingdom enterprise – taking the Gospel of Jesus Christ to unreached people in unique and innovative ways. With that purposeful call made clear, Allan's story then is one of God's timely and abundant provision. Ultimately this remarkable account is a testimony to the faithfulness of God. Which reminds us of another important biblical principle: God never calls us to a work that He does not also enable us to accomplish by the power of the Holy Spirit. Just ask Allan McGuirl."

Frank Wright, Ph.D.
National Religious Broadcasters

Printed in Canada